THE NECESSITY FOR NUCLEAR POWER

THE NECESSITY FOR NUCLEAR POWER

Geoffrey Greenhalgh

Graham & Trotman Limited

First published in 1980 by

Graham and Trotman Limited
Bond Street House
14 Clifford Street
London W1X 1RD

Softcover reprint of the hardcover 1st edition 1980

ISBN-13: 978-0-86010-249-6 e-ISBN-13: 978-94-011-7350-6
DOI: 10.1007/978-94-011-7350-6

Typeset in Great Britain by
Supreme Litho Typesetting, Ilford, Essex.

Contents

List of Figures

List of Tables

Foreword

Energy is no longer a purely technical and commercial question; it has become a political issue affecting the welfare of all mankind with far-reaching implications for the preservation of world peace. It is therefore vitally important for all of us that the right energy decisions be taken without delay and that the important contribution which nuclear energy can make toward the solution of the world energy problem should not be overlooked or, even worse, discarded.

It is now recognized that the only significant choices we have, until at least the end of this century, for the production of electric power are coal and nuclear energy and at some places hydropower. Of course, we have to use all other alternative energy sources available and capable of development, but one should realize that by the end of the century, those sources can only make a marginal contribution. The shrinking world reserves and rising costs of petroleum will eventually eliminate it as a source of energy, except for propulsion purposes and uses by the petro-chemical industry. Conservation measures in the more affluent countries and higher priority given to alternative energy technologies may only retard the growth of the demand for electricity.

To the developing world also the problems facing the nuclear industry cannot be a question of indifference. The less nuclear power that the industrial countries generate, the more oil they will consume, the higher the price will go, and less oil will be available to the developing world. The same is true of coal. Moreover, a healthy nuclear industry in the industrial countries is an indispensable basis for the expansion of nuclear power in developing countries themselves.

What is the reality behind the public's apprehensions about nuclear power? By the end of 1979, the world's installed capacity of nuclear power amounted to 120,000 MW, or six percent of the world's generating capacity. By 1985, on the basis of plants now being built, the share of nuclear power will increase to about 16 percent of the electricity annually produced. In 1985, it will represent the equivalent of more than

400 million tons of oil a year. For comparison, in 1978 Saudi Arabia produced 420 million tons of oil. Obviously, the contribution of nuclear energy is one we cannot afford to do without.

If there is a danger to mankind it does not lie in the use of nuclear energy to generate electric power. It lies in nuclear armaments and in the risks of proliferation; it could lie in the political tensions which may follow the failure to develop nuclear power and other energy resources as replacements to declining petroleum resources.

Dr Sigvard Eklund
Director-General,
International Atomic
Energy Agency, Vienna

1
The need for more energy

Energy supply is the basis on which the economic system rests. If it is believed that supplies are limited it would require the adoption of new methods of manufacture and production, of transport, and probably new systems of government to adapt to an energy restricted future. If on the other hand it is believed that, by accepting all new technologies there need be no fundamental restriction on future energy supplies it is possible to consider and plan for a future economic growth which will enable the large differences between rich and poor to be reduced – both within and between nations. This goal is only likely to be achieved by distributing shares of an evergrowing stock of wealth rather than by a forced or even voluntary sharing out of a fixed amount of resources. In this way it would be possible to remove potential sources of national or international conflict.

Yet the relationship between energy supply and demand is not a simple one. To a large extent the availability of supply creates its new demands and removes constraints that would otherwise have impeded the rate and diversity of economic growth and development. This has been pointed out by S. Schnurr of the Centre for Energy Policy Research:[1]

> "As late as 1870, about three-quarters of all the energy used in the United States was still coming from fuel wood, but the transition to coal was under way and coal soon became the dominant source. What was of primary significance in this transition was not that coal could substitute for wood in existing energy uses but rather that this was a change from a fuel resource severely limited in supply to another available in apparently endless amounts. The use of coal thus opened the way for the large-scale, unimpeded growth of iron and steel production. Ample supplies of iron and steel, in turn, made it possible to build and operate a railroad network that blanketed the country and to produce the machines required for the expansion of manufacturing. Once the fuel constraint was broken, one development led to another – a dynamic sequence that laid the foundation for modern industrial society."

In a similar way with the expansion of liquid fuels and of electricity other constraints to economic growth were overcome. Electricity and the electric motor made possible productivity increases in manufacturing by replacing power drives by shafts and belting from the in-house prime mover. With liquid hydrocarbons, the tractor and synthetic fertilisers enabled agricultural production to increase beyond the limits set by natural fertilisers, animal draft power and hard manual labour. As agricultural productivity increased, farm workers became available to take up other work. Geographical constraints were removed by the railways, motor vehicles and aircraft which have enabled raw materials, finished goods and people to move freely, and with air conditioning previously hostile environments with extremes of heat or cold can now be more readily settled and developed.

While it can be demonstrated that ample supplies of energy have enabled economic growth to occur there is now a danger that limitations of energy supply will become a constraint on economic growth: "It is very worrisome to contemplate a future in which energy supply – the constraint breaker par excellence in the past – becomes a constraint itself."[1] To assess this an attempt must be made to balance forecasts of energy demand and supply, in both the short term, up to the turn of the century, and the longer term.

Despite the growing awareness of the difficulties over future oil supplies it is widely believed that world energy demands will continue to grow. The driving forces are an increasing world population combined with an increase in gross domestic product (GDP) for both the developing and industrialised countries.

Economic growth has, however, now become a controversial subject. It can be regarded as the root from which arguments over the use of nuclear power grow. The anti-growth philosophy follows two main themes. The first is that economic growth is materialistic, philistine, polluting, that it panders to the artificially created needs of the "consumer society", that it destroys aesthetic values, leads to a deterioration in the "quality of life" and is some how basically immoral. The second theme is that economic growth cannot continue since it must, inevitably, be halted by the limits of the available resources which can be extracted from the environment, or by the limits of the environment to accept the discharge of pollutants which are said to be a necessary concomitant with growth. Although economic growth, as measured by increases in GNP has been attacked as not being a measure of total welfare, of culture or even happiness, these are largely subjective value judgements. Where attempts have been made to compile a Physical Quality of Life Index (PQLI) by combining literacy rates, life expectancy and infant mortality and ranking them as a scale from 0–100 it can be seen that although there are some variations within the individual groups of countries listed in Table 1 the average PQLI does rise steadily with the per capita GNP for the lower, lower middle, upper middle and high groups.[2]

Table 1: *GNP and the physical quality of life*[2]

National income	*Average per capita GNP (dollars)*	*PQLI*[a] *achievement*
Lower	152	39
India	140	41
Kerala, India	110	69
Sri Lanka	130	83
Lower middle	338	59
Malaysia	680	59
Republic of Korea	480	80
Cuba	640	86
Upper middle	1,091	67
Gabon	1,960	21
Iran	1,250	38
Algeria	710	42
Taiwan (Republic of China)	810	88
High	4,361	95
Kuwait	11,770	76
United States	6,670	96
Netherlands	5,250	99

[a]Physical quality of life index.

There can also be little doubt that increasing economic growth has been and is the only means of providing the large majority of the world's population with those goods and services which were once the prerogative of the few. Any alternative may lead to the introduction of a dirigistic economy with an associated police state, or at the international level a conflict between the underdeveloped and the industrialised countries. As Professor Beckerman has said in his important book *In Defence of Economic Growth*:[3]

> "Only an altogether unparalleled optimism can lead one to believe that the vast mass of the population will voluntarily accept an abandonment of the goal of economic growth, at least for the foreseeable future. This means that if growth were to be abandoned as an objective of policy, democracy too would have to be abandoned . . . the costs of deliberate non-growth, in terms of the political and social transformation that would be required in society are astronomical."

The second argument against continued growth, typified by the "limits to growth" school can be countered by the expansionist view put forward, for instance, by Chauncey Starr in his article "The Growth of Limits".[4]

> " . . . Science and technology historically have opened new frontiers for mankind, not only permitting but also stimulating the 'growth of limits'. I do not accept the premise that constraints on such growth are in view or that our long-range planning horizons should be determined by today's perceptions of existing limits . . . "

> "The recently popular 'limits to growth' theme was based on the assumption that natural resources and ecology impose a near-term limit on our ability to continue to derive materials from the earth and to handle safely residuals of growing industrial processes. For proponents of this theme the obvious course would be to limit the personal economic expectation of those now alive and to limit sharply the growth in the world's population. This is a seductive dogma . . . but what about the poor people of the world, both those within the industrialised nations and the vast numbers in countries that have not yet caught up with the modern world? For them limits to growth would mean lives of hopelessness and despair."

The argument about the desirability or otherwise of growth still continues, but in the meantime life goes on and practical men must try to plan the future. The World Bank, in their *World Development Report*,[5] August 1979, have published figures for the future growth rates of the Developing and Industrialised countries up to 1990. These are cautiously described as projections, not as targets for international decision making or as precise forecasts for the future. They do however indicate an expectation that economic growth will continue (see Tables 2 and 3).

Table 2: *Developing countries: growth of gross domestic product, 1970–90*
(Average annual percentage growth rates, at 1975 prices)

	1970–76	*1977*[a]	*1978*[a]	*1975–85*	*1985–90*
Low income countries	3.4	5.7	5.4	4.7	4.9
Africa	2.6	4.0	3.4	3.7	3.8
Asia	3.5	6.0	5.7	4.9	5.1
Middle income countries	6.2	4.6	5.0	5.3	5.8
All developing countries	5.7	4.8	5.1	5.2	5.6

[a]Estimates based on preliminary and incomplete data. Source Ref. 5.

As long as economic growth is related to activities which involve the transformation and/or the transport of matter, there must always be a correlation between economic growth and increasing energy consumption. It is not possible to manufacture goods or to market them without using energy. In addition, energy is also required to provide civilised societies with an environment more suited to their desires than the climate that nature provides. This will require energy for heating or cooling. Any increase in personal

Table 3: *Industrialised countries: growth of gross domestic product, 1960–90*
(Average annual percentage growth rates, at 1975 prices)

	1960–70	*1970–78*[a]	*1970–80*	*1980–90*
North America	4.0	3.4	3.3	4.0
Japan and Oceania	9.4	5.1	5.1	5.9
Western and Northern Europe	4.7	2.8	2.9	3.8
All industrialised countries	4.9	3.4	3.4	4.2

[a]Estimates for 1978 are based on preliminary data. Source Ref. 5.

comfort will consume energy. And as over recent years the more hostile regions are becoming increasingly colonised, Alaska, Siberia at one extreme and desert settlements at the other, this energy requirement will increase. Improvements in environmental quality with stricter standards on the discharge of pollutants, cleaner air and water, lower noise levels etc. will also require the expenditure of more energy than if controls on such discharges did not exist.

The relationship between economic growth and energy consumption is a complex one. It can vary quite widely even among the industrial countries and depends on many factors, of which climate, life-style, level and nature of industrial and agricultural production are some of the more obvious. Industrial production is one of the largest single factors and can account for one third of GDP and here there is a wide range in the energy intensities of different industrial sectors. With a factor of over 10 between an energy intensive industry such as iron and steel compared with the food, drink and tobacco industries.

As the contribution these different industries make to the national economy increases or declines the energy consumption GDP ratio will also change.

With the expected move towards "the post industrial economy" the higher energy consuming industries such as iron and steel, chemicals etc. will play a lesser part in the present highly developed countries which will show faster growth in the services sector. But this merely pushes the problem of the energy intensive industries onto the new developing countries, such as Korea and Brazil. It does not mean that the world will use less energy. The recent changes in growth/energy ratio for a number of countries are shown in Figure 1. The high figures for the UK reflect the habits of a country whose industries and way of life have been built up on cheap fuel; the decline of heavy industry, the growth

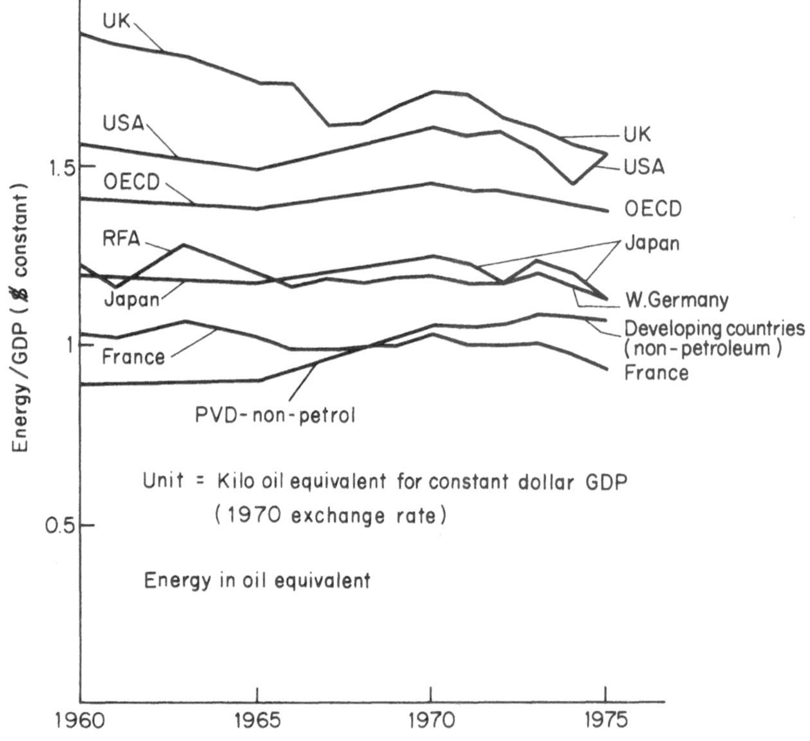

Figure 1: *Recent changes in growth/energy ratios for a number of countries. (Source: Ref. 6.)*

of the service sector* and a more frugal regard for higher price fuel is now bringing about a steady reduction. The USA is another country where consumption habits were formed at a time of fuel self-sufficiency, whereas France which has over this period relied heavily on fuel imports, 35% in 1960 rising to almost 75% by 1975, has a consistently lower energy/GDP ratio than the UK and USA.

For the future it has been suggested that the average ratio, now about 1, will fall to 0.8 by the year 2000 and about 0.6 by 2025. But there will be a distinction between the developing and the industrial countries. As the latter move to the service based economy of the post industrial society their energy/growth ratio

*For the UK total employment in manufacturing, mining and farming dropped by 2.04 million to 8.42 m between 1961 and 1968, while during the same period employment in other services – which covers most of the public sector as well as the professions and finance – rose by 2.9 m to 12.66 m. Central Statistical Office: Economic Trends December 1979.

may fall to 0.5–0.6 while in the developing countries which will increasingly take over the responsibility for heavy industry, chemicals, fertilisers, primary metals, heavy engineering, shipbuilding etc. the ratio could remain at a higher level of above 1.0.

This so-called decoupling of economic growth from energy requirement, i.e. a fall in the ratio below the value of 1, is regarded in some quarters as having an almost moral significance. But the change can only come by (1) reducing the volume of energy-requiring activities such as industrial production, (civil servants, lawyers, bankers etc. use less energy than miners and metal workers); (2) increasing the efficiency of energy production; (3) energy conservation.

All the three factors are exhaustible so that a full decoupling of economic growth from energy can never be achieved. The degree to which the high energy consuming industries can be replaced by low energy service type activity must be limited and any reduction of industrial activity within one society is likely to be reflected as an increase in another. The requirement for manufactured goods and products can only increase as living standards rise across the world. A decline of energy intensive industry in the West will be co-incident with the rise of such industries in the new industrial countries. With the high cost of imported energy, the Japanese aluminium smelter capacity is to be reduced in size by one-third, and aluminium will be imported into Japan; it is expected that the European aluminium industry will be similarly affected. There is however now a rush to build aluminium smelters in Australia, using cheap, plentiful Australian coal and production is now starting in the Middle East oil countries. The steel industry in UK, Europe and USA is also facing increasing difficulties and production is declining, whereas production is rising in Brazil and South Korea. Brazil is also emerging as a major shipbuilding country. The post-industrial society is likely to make more time available for leisure and cultural activities which could themselves be energy consuming; there is no reason to believe that this will necessarily be a low energy society.

The second factor, efficiency of energy production is limited by thermodynamic laws as well as by technical and cost considerations.

The third factor, energy conservation, is more fundamental, and indeed plays an essential part in efficiency of energy production and use. It is one of the main planks on which the structure to support a balance between energy demand and supply must rest. As the Ford Foundation in their report *Nuclear Power Issues and Choices* noted in 1975:

"the fourfold increase in the price of imported oil since 1973 justifies conservation investments today that would not previously have repaid their cost"

and further increases in the price of energy will provide a continuing economic driving force for greater conservation measures. There is room, particularly in countries such as the UK and USA where the energy consumption per unit of gross national product (GNP) is high, for the introduction of more efficient energy processes in industry, and where domestic consumption habits and standards were formed in an age of cheap fuel. It has indeed been argued that for both the UK and USA, conservation alone could be sufficient to solve the future energy supply problem. For the USA the Demand and Conservation Panel of the National Research Council Committee on Nuclear and Alternative Energy Systems concluded in April 1978 that: given enough time to respond to prices, regulations and incentives, US energy demand could grow at a much slower rate than in the recent past and might even fall while still assuming that the GNP would increase twofold. For the year 2010, this study found that low growth energy futures could plausibly range from a consumption of 63Q to 137Q compared with the actual figure of 76Q in 1977 (1Q = 10^{15} Btu). The low figure assumes very aggressive conservation policies, the second a continuation of present policies. For both figures the Gross National Product was assumed to be about double that of 1977 (in constant dollars).

A similar study for the UK has been carried out by the International Institute for the Environment and Development on a grant from the Ford Foundation. Their report *A Low Energy Strategy for the UK* which looks ahead to the year 2025 presupposes that energy savings can continue to be made at a rate above the rate of energy increase that would be the consequence of an increase in GDP and argues that it would then be possible to maintain economic growth in the UK – at an annual rate 3% GDP during the 1980s and 2.4% for the 1990s – without any increase in primary energy use by the year 2000, and even a fall of 8% by the year 2025. With lower GDP growth rates of 2.5% for the 1980s and 2% for the 1990s primary energy use would fall 7% by 2000 and 22% by 2025.

Many of the proposals made to bring about this reduction in energy use are sensible, desirable, and will undoubtedly be introduced on an increasing scale. The weakness is to assume that all will be put into practice and that all will work as efficiently as is claimed – often for technologies at the earliest stage of development or even as yet only theoretical possibilities.

Although conservation is an essential and important constituent of any energy policy there is a danger that it is being regarded as an end in itself; that a low energy society is in some way more moral.

While the need for greater conservation cannot be too strongly pressed, such claims underestimate the practical difficulties and overestimate the contribution that restraint by the industrial countries might make to easing the energy problems of the developing world.

The difficulties of reducing energy consumption in a rich country are shown by the experience of "Project Independence" launched in 1973 by President Nixon with the objective of achieving a total energy independence for the US by 1980. Even though, following a realistic assessment in 1974 by the newly created Federal Energy Administration, the date for energy independence was put back to 1985, the chances of this being achieved are low. Despite the gains in domestic oil supply from Alaska, US oil imports in 1977 were 20% above 1976 and continue to rise. The US Gross Energy Consumption shows a continuing increase (Figure 2).[2]

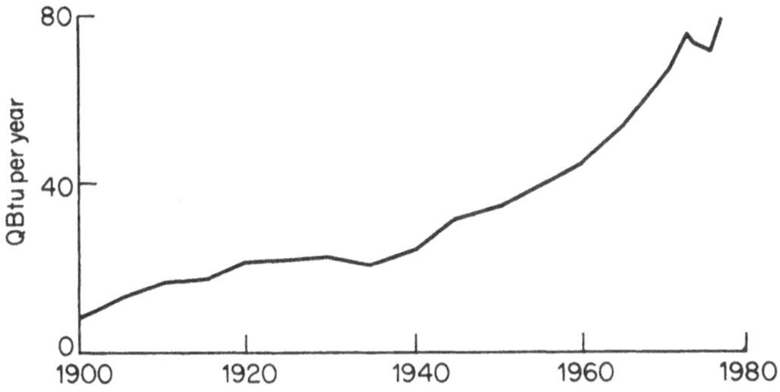

Figure 2: *US gross energy consumption, 1900–1977. (Source: Ref. 2)*

The short lived downward movement 1974–76 more probably reflects the effects of the recession which followed the oil price rise rather than the effect of a growing response to conservation measures. The upward rise of the curve has recommenced as steeply as ever. The same effect can be seen in the US commercial and residential usage of energy (see Figures 3 and 4). In both of these curves the increasing share of electricity is a point of interest.[2]

On the other hand there are signs that the US industry is increasing its energy efficiency. In 1977 a 5.6% increase

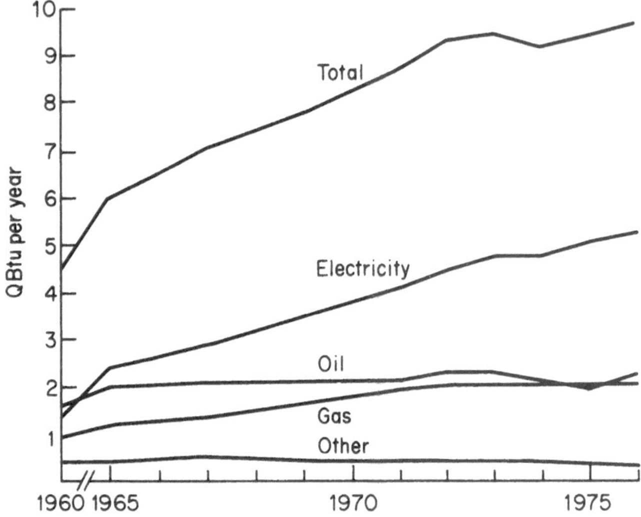

Figure 3: *US commercial energy use, 1960–1976. (Source: Ref. 2)*

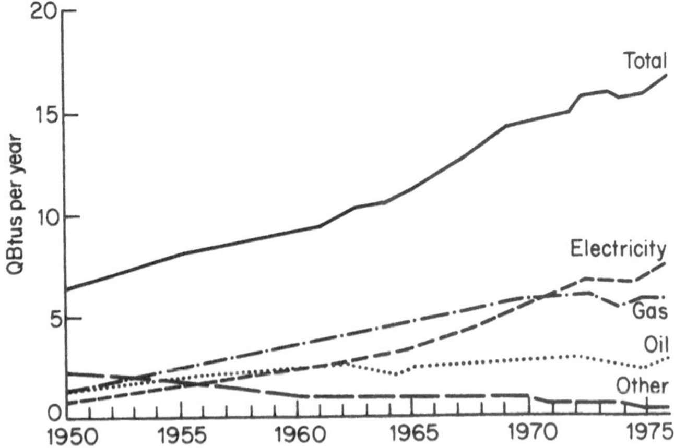

Figure 4: *US residential energy use, 1950–1976. (Source: Ref. 2)*

in production was obtained with a 0.1% reduction in energy use compared with 1976. It is claimed that this is a response to increased fuel prices and concern over the availability of some fuels.* Such a response will eventually come in the commercial and residential sector, but this will be slower as it depends on the individual decisions of the very much larger numbers of commercial and residential users compared with the more swift action that can be taken by the much smaller number of large energy using establishments in industry.

Energy conservation is however an option which can only be exercised by the richer countries which are the large energy consumers. But these represent only a minor fraction of the world's population. The inequality whereby 16.2% of the world's population in the industrialised countries consume 57.3% of the commercial energy production is well known, but the figures are even more striking when presented graphically (Figure 5).[7]

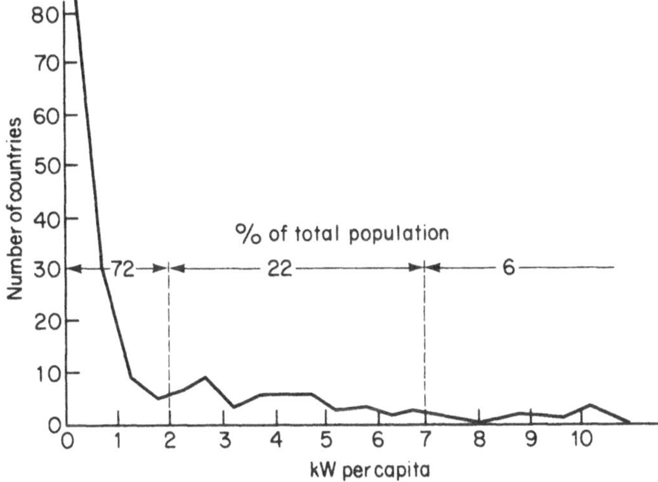

Figure 5: *Distribution of per capita energy consumption in 1971*

72% use less than 2 kW per capita and of these a significant number as little as 0.2 kW per capita, whereas at the other extreme 6% of the world population has an energy consumption of more than 7 kW per capita of which the highest is the USA with 11 kW per capita.

*It is however more likely to be due to a decline in the energy intensive primary raw material industries with a move to secondary manufacturing industries of much higher added value with a lower energy requirement.

A major problem will then be in finding the energy resources to fuel the economic development of the poorer countries. Table 4[5] from World Bank statistics shows that there is a correlation between GNP and a number of factors which can be taken as indications of social and physical well being, such as infant mortality rates, life expectancy, protein consumption, availability of medical services and literacy.

Table 4: *Selected social indicators for groups of nations, 1970*

Per capita GNP group	*Infant mortality (per thousand live births)*	*Life expectancy at birth (years)*	*Average per capita daily protein consumption*	*Population per physician*	*Adult literacy (percentage)*
$100 and below (18 nations)	159.5[a]	43.7	56.5	21,820.5	NA[b]
$101–200 (18 nations)	113.1	46.5	57.7	8,878.5	NA[c]
$201–375 (26 nations)	100.2	55.9	59.3	3,436.5	43.1
$376–1,000 (26 nations)	55.6	61.4	68.2	1,729.4	70.3
$1,001 and above (30 nations)	19.1	71.6	89.4	715.6	97.6

[a]1970 figures not available; 1960 figures used instead.
[b]Limited available national statistics range from 5 percent for Upper Volta and Somalia to 56 percent for Indonesia.
[c]Limited available national statistics range from 15 percent for Sierra Leone to 85 percent for Sri Lanka.
(Source: Ref. 5)

The World Bank has also estimated that even with increasing economic growth rates of over 5% per year, some 600 million people, about 10% of the world's population will be living in absolute poverty by the year 2000 (Table 5).[5] *

The poverty of the poorest countries is self-defeating. People living at the margin cannot alleviate their present needs by energy conservation, whilst conserving for the future is a luxury they cannot afford. They must, inevitably, abuse their surrounding physical environment. Wood taken for fuel and heat accelerates deforestation which in turn leads to erosion, flooding and a

*Their number is reduced with the high growth, and increased with the low growth scenarios.

Table 5: *Levels of absolute poverty under alternative scenarios, year 2000*

	Percentage of population	*Millions of absolute poor*
Base scenario		
Low income countries	22	440
Middle income countries	10	160
All developing countries	17	600
High scenario		
Low income countries	17	340
Middle income countries	8	130
All developing countries	13	470
Low scenario		
Low income countries	26	520
Middle income countries	12	190
All developing countries	20	710

Source: Ref. 5.

decline in soil fertility. As firewood becomes scarce and expensive, people turn to substitutes such as burning dried animal dung, as in the rural areas of India, which robs the soil of a valuable natural fertiliser and structural stabiliser, thus further depleting the fertility of the land. For urban dwellers the usual substitute is kerosene, but this is now becoming an increasingly expensive fuel and the poorer populations are forced back to firewood or wood charcoal. The consequences for urban settlements can be disastrous. An extreme example, quoted in the 9th Annual report of the US Council on Environmental Quality, can be found at Ouagadougou in Upper Volta where the town is surrounded by an utterly denuded circle some 70 km in diameter within which all trees have been consumed for firewood. As the distance that the fuel must be carried grows the cost rises, and it is said that the average Ouagadougon labourer spends 20% of his annual income for fuel wood. By contrast the average US family spent only 6% of its income to meet all its energy needs in 1973.

The difficulties of obtaining fuel are then worst in the towns and cities of the poorest countries: but this is where the population growth is now fastest. Massive migrations to urban centres accelerate as rural employment fails to keep pace with population growth and industrial jobs pay higher wages. In the poorest countries – those with per capita GNP less than $100 – while the population grew overall at an average rate of 2.3% in 1970, the growth of urban population was nearly twice as rapid at 4.4%. Countries with slightly higher GNP levels experienced even higher urban growth rates of up to 5.4% per year.[8] Overcrowded shanty towns and slum tenements are growing in cities like Bombay, Jakarta, Karachi and Manila, which will, it is estimated by the World Bank, have populations in excess of those of New York, Tokyo and other industrial centres by the year 1990. This transfer of people involves very large numbers. Between 1950 and 1975 the urban areas of developing countries absorbed some 400 million people: between 1975 and 2000 the increase will be close to one billion people (Figure 6).[5] This is not to argue that this

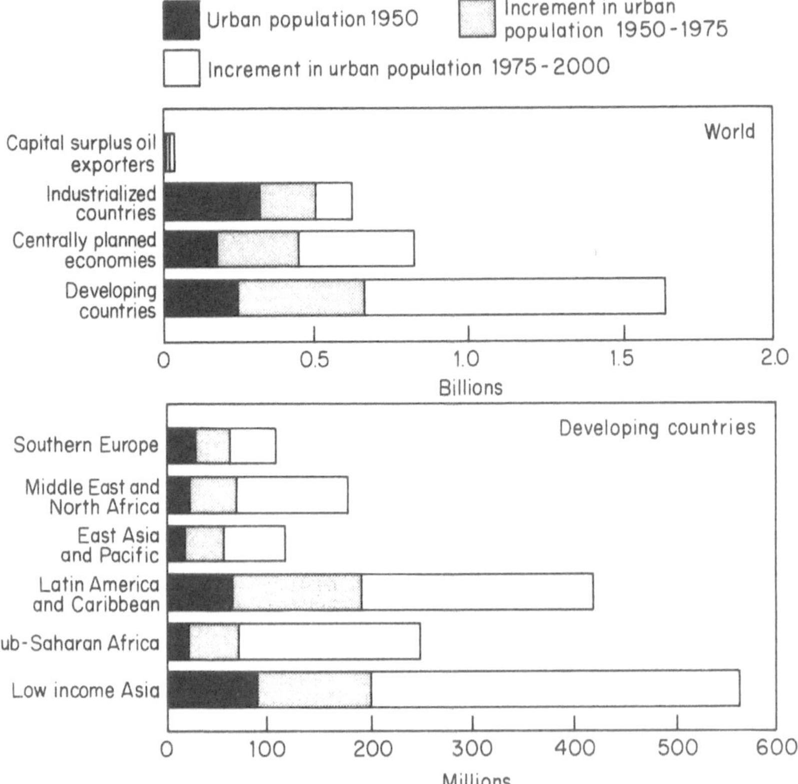

Figure 6: *Urban population estimates and projections, 1950–2000. (Source: Ref. 5)*

urbanisation is desirable, but in the short term at least it seems unavoidable. Conditions could however be made more tolerable with adequate supplies of energy. But urbanisation has an immediate effect on energy requirements. Urbanised people in the developing countries are known[9] to use almost 10 times more energy per capita than those in the rural areas.

This rapid transfer of population brings other problems. For the industrialised countries urbanisation took place slowly over many decades so that the economic, social and political institutions and structures were able to evolve. In the developing countries this change is occurring much more rapidly, with lower incomes and less opportunity to alleviate the problems by international migration. These developing cities then suffer more severely from air and water pollution. A contributory factor to the air pollution is the closely packed housing where firewood, kerosene or coal are burned for cooking and heating. Water pollution comes from inadequate or even non-existent sewage disposal systems as well as industrial effluent. The World Health Organisation reports that in 1975, 1.2 billion people (almost two-thirds of the population of the developing countries) lacked access to clean water supplies and 1.4 billion lived without even simple waste disposal systems. "As a result", a World Watch Institute report stated, "environmental pollution by untreated human waste exacts a disease and death toll that dwarfs the known toll of industrial pollutants".[10] The World Bank estimates that water delivery systems can range in cost between $1 per person for simple shallow wells up to $50 per person for piped distribution. Sewage systems cost between $1 per person for a pit latrine up to $200 per person for a sewage disposal and treatment system. The goal of providing "clean water for all by 1990" agreed upon at the 1976 UN Conference on Human Settlements and reaffirmed by the 1977 UN Water conference is estimated to cost as much as $125 billion.[11] It will also add vastly to the energy demand of the developing countries.

These developing countries rely heavily on energy from firewood, charcoal, plant and animal residues, and to a lesser extent on wind, water power and solar energy. Although such forms of energy account for only about 5% of world energy consumption they provide about half of the energy for the developing countries, and more than 85% in the rural areas. In low income countries such as Mali, Nepal and Tanzania over 90% of the energy used comes from these sources, principally for domestic use since there is little industry. Programmes for afforestation or reforestation, especially with quick growing varieties of trees, can provide some relief of energy supplies in rural areas – perhaps the only relief since rural electrification is not practicable in areas of low density of population – but the high bulk, low energy concentration of such fuels makes them impracticable for wide use in the large urban conglomerates. But it is here that central generation of electricity would not only be feasible but desirable.

The reverse process of urban to rural movement which many

may regard as desirable is the luxury of the rich; the hitherto universal pattern of rural to urban migration is only now being reversed in the high energy s 'ciety of USA and could be regarded as an indication of the movement towards a "post industrial" society. The peak of rural-to-urban migration within the USA had passed by the mid-1960s. Although most metropolitan areas are still growing, an increasing number of the large USA cities are showing population declines with an outward migration to the suburbs and rural areas.[2]

A second factor in determining the world energy requirement is the growth in population. Although there are now hopes that an end can be foreseen to the exponential growth in world population of the last century, it may take a further 150 years before the population levels out at a figure of around 10–15 billion people compared with the present figure of 4 billion (Figure 7).[7]

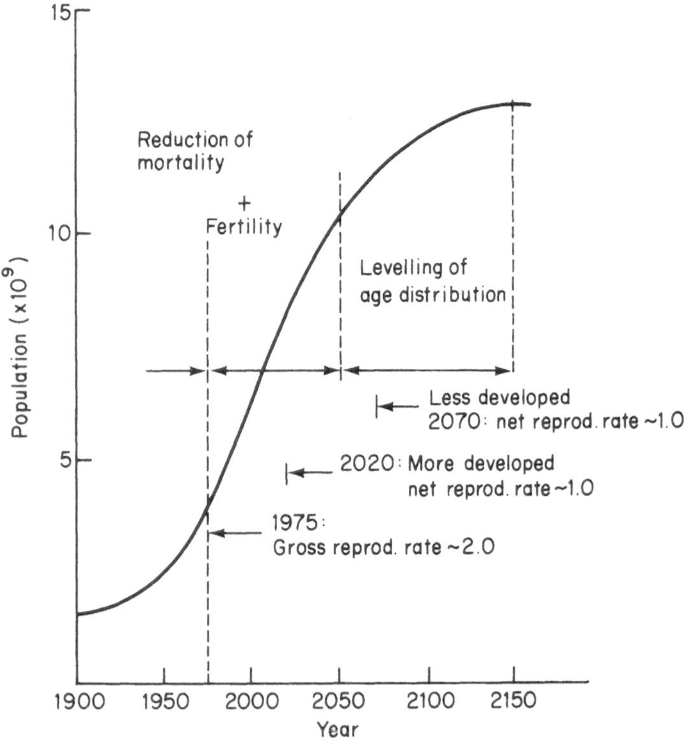

Figure 7: *World population growth. (Source: Ref. 7)*

This curve from the UN World Population Conference in

Bucharest, 1974, shows an increase from 1.6 billion in 1900 to 4 billion in 1975 which is largely due to a reduction in mortality rate, starting in the developed countries and spreading to the developing countries. In a similar way it is believed that the reduction of birth rate now being observed in some developed countries will also spread to the developing countries. While the population growth rate of the industrialised countries is projected at 0.4% per year, in many of the developing countries growth rates of over 2% per year are expected. Declining birth rate is associated with increased urbanisation, improved education and more extensive participation of women in the labour force, but a major factor is the increasing use of contraception. The potential for this is illustrated in Table 6[5] which shows the relationship between contraceptive use and birth rates in some developing countries. Under optimistic assumptions, which include a favourable economic development of the third world, the UN report

Table 6: *Contraceptive use and crude birth rates in selected developing countries, 1977*

	Percentage of married women of child-bearing age using contraceptives[a]	*Crude birth rate per thousand population*
Kenya	4	51
Pakistan	6	45
Indonesia	19	37
Mexico	21	38
Egypt	21	36
Thailand	32	32
Malaysia	34	29
Sri Lanka	44	26
Colombia	49	30
Hong Kong	64	19

[a]These data refer to the latest available information covering the period 1975–1977. (Source: Ref. 5)

forecasts that a net reproduction rate* of 1.0 will be reached in the developed countries by 2020 and in the developing countries by 2070. This would give a world population of about 10 billion people by the year 2050. A further increase of some 3 billion can be expected as a consequence of the levelling off of age distribution as the mortality rate continues to decline.

*A net reproduction rate of 1.0 means that on average each couple will produce 2 children who will survive until their reproductive age, thus maintaining an equilibrium state.

Population growth will, however, not be uniform. As already noted the fall in fertility is expected to start in the developed countries, while the developing countries follow later. This means that population growth will be greatest in the present developing countries. The result expected is shown in Figure 8.

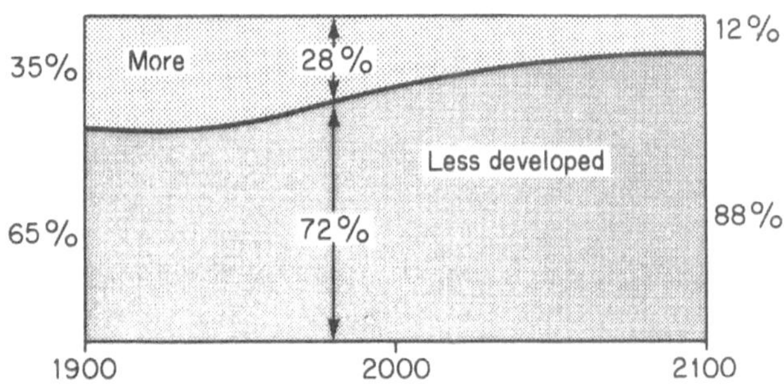

Figure 8: *Changes between more and less developed population shares. (Source: Ref. 7)*

It is these less developed countries which will experience the greater rise in energy use above their present low levels. So even if the industrial countries were able to slow down their increase in energy or hold to the existing use this would have little effect on the increase in world demand.

Feeding such an expanding population, could, in some parts of the world present a problem even in the short term. The United Nations World Food Council has declared[12] that the problem of hunger is "getting worse rather than better", and crisis is now threatening unless nations move now to avert it. Because of inadequate investment in food production in developing countries, world food consumption in 1980 would outstrip production. A report by the Food and Agriculture Organisation to a UN food conference, 12 November, 1979, says that with rising cost of imports caused by the increasing price of energy poor countries are becoming increasingly unable to feed themselves. Grain imports of the developing countries have grown rapidly in the four years from 51 million tons in 1975 to about 80 million tons in 1979. The report concludes that the "Third World" faces the prospect of increasing shortages of food that could spell economic disaster, hunger and malnutrition on a massive scale by the turn of the century.[13] Figure 9 illustrates the results of a survey,[14] and

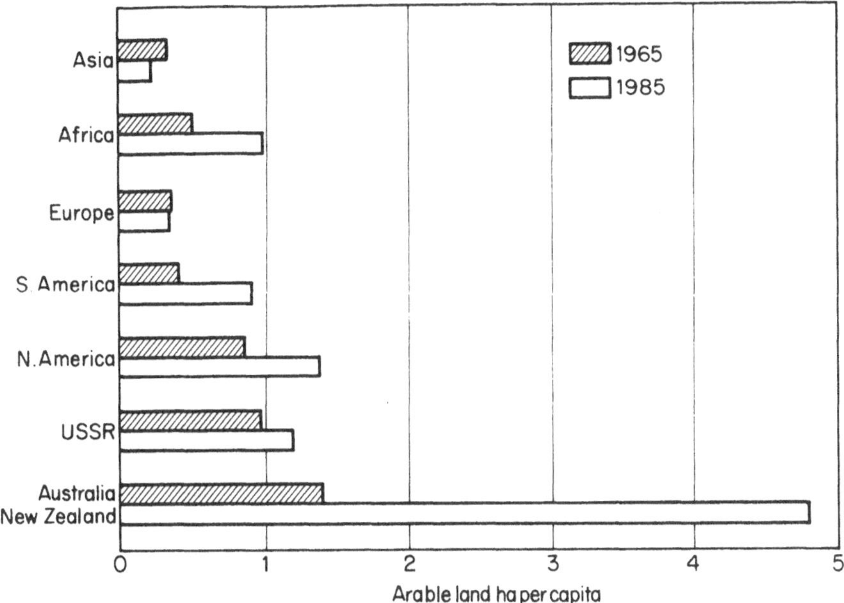

Figure 9: *Arable land per capita; a survey (1965) and an estimate (1985)*

future estimates of arable land per capita in different regions. It shows that between 1965 and 1985 the arable land available per capita in Europe and S.E. Asia will diminish to about 0.5–0.3 hectares per capita. A sufficient food output can then only be obtained by intensive production methods which require a higher input of energy. The growth of US energy requirements per calorie of food output (see Figure 10) increasing tenfold from 1910 to 1970 indicates the trend that will be followed in other parts of the world as intensive food production increases.[14]

In the future some relief will come from the production of synthetic protein from natural gas and oil with the product being fed through an animal system or directly for human consumption. We should not then consider burning hydrocarbons; we may need to eat them.

World population growth will eventually level out with the development and wider acceptance of methods of fertility control, and perhaps even more fundamentally by providing incentives to use these control methods either through education or by enabling

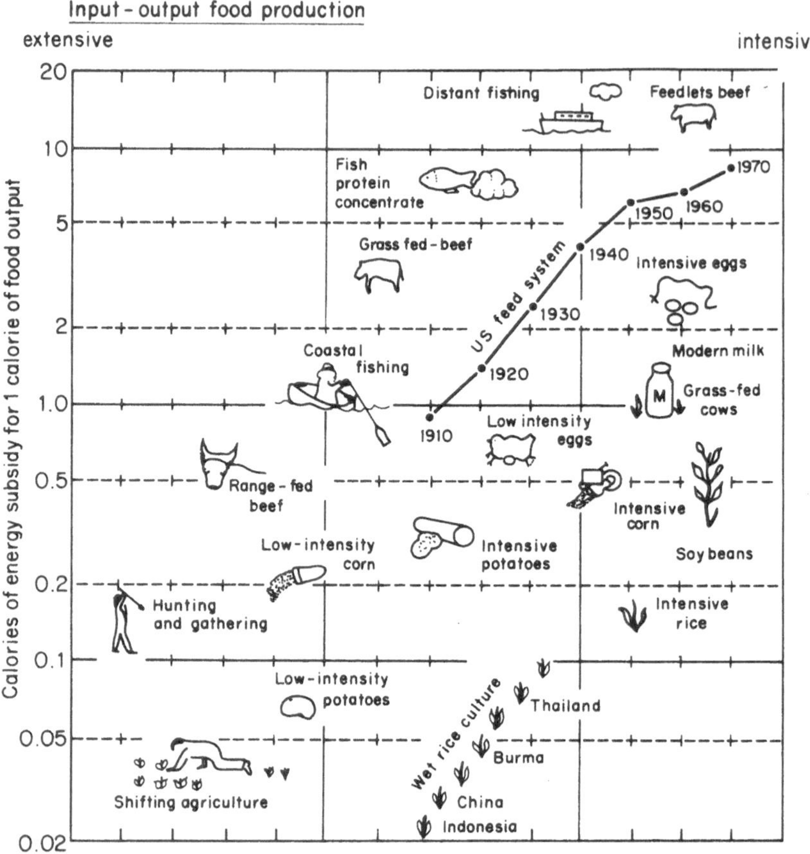

Figure 10: *Energy subsidies for various food crops. The energy history of the US food system is shown for comparison.*

parents to acquire sufficient material wealth for their old age, so that they can be confident that they and their family unit can be maintained by the survival of only two children. This applies in particular to those developing countries which are predominantly of a peasant based agricultural society and where the population is increasingly at the highest rate. It could be that the provision of an adequate old age pension may be one of the most effective ways of slowing down the growth of population, but this will require a wealthy economically advanced society. Weinburg and Hammond[15] in 1971 considered it very likely that world population could, before it levels out, ultimately reach 10 billion, that 15 billion is quite likely and 20 billion not impossible. They take the mean figure of 15 billion not because they recommend it as desirable, but because they see no acceptable humane way to prevent it. By the time, perhaps 100 years from now, when a stable population has been reached the availability of adequate resources of energy will be a critical factor. A world population of that size could only be fed with an increased food supply. This will require a high energy input into fertiliser production, water control, irrigation and the use of agricultural machinery. Energy can convert sea water, air, and desert land into hydrogen, fertilisers, fresh water, foods and industrial goods. An adequate old age pension, a base premise for a stable population, requires a high minimum level of wealth in society and hence a high minimum energy use. The material requiements of 15 billion people at a high standard of living lead to concern as to the possible exhaustion of essential resources, but provided sufficient energy is available for the necessary extractive work it seems probable that man's requirements can be met for a long time to come. Some substitutions, adjustments and compromises will be inevitable, and there will be expensive recycling of scarce substances, but the key requirements appear to be available.

In estimating the future energy requirement Weinberg and Hammond start from the present USA consumption of energy in all forms at a rate of 10 kW(th)* years per year and assume that the steady state demand for a 15 billion world population will be twice this figure, with the whole world being brought up to the same level. This is on the basis shown in Table 7.

Table 7: *Energy budget per capita for a steady-state civilisation (kilowatts fuel equivalent)*

Present US level (1970 figure)		10.0
Adjustments for the future		
Steel, aluminium, and magnesium production	0.1	
Recovery and recycle of scarce elements (copper, zinc, tin, lead, mercury, gold, silver, titanium, etc.)	2.0	
Electrolytic hydrogen	2.5	
Water by desalting (100 gal/day)	0.3	
Water transport to cities	0.1	
Air conditioning to cities	0.3	
Intensive food production	0.2	
Sewage and waste treatment	0.5	
Total adjustments		6.0
Contingency		4.0
Total budget		20.0

*th = thermal

Table 8: *Growth and levels of gross domestic product under alternative scenarios*

	Average annual growth rates, 1980–90 (percent, at 1975 prices)					
	Gross domestic product			*Gross domestic product per capita*		
	Base	*High*	*Low*	*Base*	*High*	*Low*
Low income countries	4.9	5.9	4.3	2.7	3.5	2.0
Africa	3.8	4.8	3.6	1.0	1.9	0.7
Asia	5.0	6.0	4.4	2.8	3.8	2.2
Middle income countries	5.8	6.8	4.9	3.4	4.3	2.4
East Asia and Pacific	7.6	9.3	6.4	5.6	7.1	4.3
Latin America and Caribbean	5.7	6.5	4.6	3.2	3.9	2.1
Middle East and North Africa	5.5	6.3	5.0	2.9	3.6	2.4
Sub-Saharan Africa	4.4	5.3	3.7	1.4	2.2	0.7
Southern Europe	5.4	6.5	4.7	4.2	5.2	3.4
All developing countries	5.6	6.6	4.8	3.3	4.2	2.4
Industrialised countries	4.2	4.9	3.5	3.7	4.5	3.1
Capital surplus oil exporters	5.0	6.1	4.6	2.2	3.2	1.7
Centrally planned economies[a]	4.2	..	..	3.4	..	..

	Gross domestic product per capita (1975 US dollars)			
	1975	*1990*		
		Base	*High*	*Low*
Low income countries	147	211	232	200
Africa	146	165	181	160
Asia	148	219	240	206
Middle income countries	950	1,476	1,622	1,354
East Asia and Pacific	582	1,399	1,638	1,258
Latin America and Caribbean	1,103	1,632	1,756	1,471
Middle East and North Africa	823	1,234	1,325	1,173
Sub-Saharan Africa	544	630	683	586
Southern Europe	1,808	3,122	3,463	2,907
All developing countries	499	773	849	712
Industrialised countries	5,865	9,999	10,747	9,381
Capital surplus oil exporters	6,192	8,439	9,332	8,049
Centrally planned economies[a]	2,560	4,351	..	..

[a]East European centrally planned economies only.

To achieve this very substantial advance over the next 100 years will require economic growth – and consequently energy growth – to continue in the industrialised countries as well as in the developing countries. The World Bank development report makes the point that the economic health of the industrialised countries is a key determinant of the growth prospects of the developing nations. The industrialised countries provide the principal markets for the exports of the developing countries; they supply the external capital and provide the technology for further development. Since however the population increase of the industrialised countries has fallen well below that of the developing countries, one effect of the continuation of growth will be to maintain or even widen the gap between rich and poor when expressed in terms of GDP per capita (Table 8).

It has been argued by Leon C. Martel of the Hudson Institute that there is a natural growth of growth which a nation follows as it develops: from slow to fast, and then a return to slow growth rates with the arrival of post-industrial society. The existing gap between the rich and the poor nations is both a result of this pattern and a necessity for raising the income of less-developed countries. Imposing limits on growth and redistributing the wealth of the wealthiest will not only *not* aid the poorest; it could increase their poverty. Growth is needed to raise the level of those who suffer a barely minimal existence to create the economic and social development that are prerequistes for slowing rates of population increase. "The poor will get richer because the rich get richer."[16]

The growth and demographic transitions can be depicted as a flattened S-shaped curve by plotting values of world population and per capita product against total world product, on a logarithmic scale as in Figure 11.

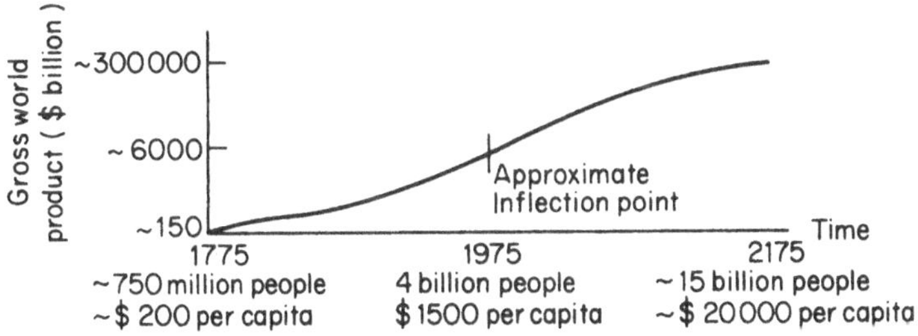

Figure 11: *Growth and demographic transitions of gross world product versus world population and per capita product (Source: Ref. 16)*

This shows all three values starting to rise 200 years ago to reach their highest rate of growth at the present time. The rates of accumulation and growth are then expected to slow over the next 200 years.

If the nations of the world were plotted along this curve in terms of their relative positions today the most developed which are now super-industrial societies evolving towards post-industrial societies,* would be seen as having passed the inflection point of population growth and being close to – or in some cases past – the inflection point of product growth. Other less developed nations would lie further down the curve with the poorest towards the bottom. And for these the rate of population growth would be high and rising and the per capita product would be low.

There is thus an economic-growth transition, analogous to a population-growth transition which accompanies and aids it.

> "Passage through this growth transition represents an enormous gain in efficiency, safety and the accumulation of surplus for both investment and consumption; populations whose nations accomplish it enjoy better health, longer lives and a vastly improved standard of living. Therefore the task today is not to limit growth, but to *prevent* the imposition of exogenous limits on it and whenever possible, but particularly in the poorest nations, to do everything that can be done to hasten it."[16]

This is a task which can only be accomplished if the world has available an adequate supply of energy.

REFERENCES

1. S. Schnurr, *EPRI Journal*, May 1978.
2. *Environmental Quality*, 9th Annual Report of the Council on Environmental Quality, December 1978.
3. W. Beckermann, *In Defence of Economic Growth*, Jonathan Cape, 1974.
4. Chauncey Starr, "The Growth of Limits", *EPRI Journal*, June 1979.
5. *World Development Report*, The World Bank, 1979.
6. Thierry de Montbrial, *L'Energie: Le Compte a Rebours*, J.-C. Lattes, Paris.
7. Second Status Report on the IIASA project on energy systems 1975, W. Häfele *et al.*, International Institute for Applied Systems Analysis, 2361, Laxenburg, Austria.
8. *World Tables 1976*, Johns Hopkins University Press, Baltimore, 1976.
9. W. Häfele and W. Sassim, *Annual Review of Energy*, Vol. 2, 1977.
10. *The Global Environment and Basic Human Needs*, US Government Printing Office, Washington, 1978, p. 16.
11. *Community Water Supply and Waste Water Disposal*, WHO, Geneva, 1976.
12. *Financial Times*, 9 October 1979.
13. *Financial Times*, 13 November 1979.
14. U. La Roche, *Energy Scenario*, Brown Boveri, Baden, Switzerland.
15. Weinberg and Hammond, Paper 240, IAEA 4th International Conference on peaceful Uses of Atomic Energy, July 1971.
16. L. Martell, *Futures*, April 1979.
17. T. Stonier, *The Times*, 13 November 1978.

*Martell defines super-industrial as the period preceding post-industrial, when large-scale industry is the major economic activity of a society. A post-industrial society meets its industrial and material needs with only a small percentage of its work force and total economic efforts. (The move towards a post-industrial society for the UK is already foreseen with reports that within 30 years Britain will need no more than 10% of its labour force to supply all its material needs.)[17]

2
Energy demand

The basic forces pushing up energy demand are population increase and economic growth. From an assessment of these it is possible to estimate future energy requirements.

This book concentrates on the period up to the year 2000 since it is over this period that the problems and difficulties that will arise from energy shortages are likely to be most acute. Given time, new energy sources could be expanded, or adjustments made in energy use – but time is in short supply.

One indication of the future growth of demand by the International Institute of Advanced Systems Analysis is based upon a population increase rising from 4 billion in 1975 to 6 billion in 2000 and eventually levelling off at something over 10 billion people (Figure 12).[1]

Above the curve of population growth are three energy growth curves. The lower one drawn parallel to the population curve shows how the energy demand would inevitably increase even if the average consumption per capita remained approximately the same as the present world figure of 2 kW (th). This "zero-growth" for the average individual would still result in the global energy demand doubling within the next 40 years before it eventually levelled out at three times the present value around the year 2100. If it is assumed that a single economic and technological life-style is slowly diffusing throughout the world this would correspond to settling at an average level equal to the present energy consumption of a country such as Spain; it would require a considerable reduction in energy use for the industrialised countries which now consume between 5–10 kW (th) per capita.

If, however, it is assumed that a modest increase in the world average energy consumption occurs, rising to 5 kW per capita, about the present level enjoyed by the UK and the Federal Republic of Germany, the energy demand would follow one of the two upper curves of Figure 12. The lower of these assumes an initial energy growth rate of 3% per year, which means that the "target" of 5 kW per capita would be achieved within 3–4 generations around the year 2060. The upper curve shows that with a faster initial growth rate of 4.5% per year the 5 kW per capita level would be reached in 2 generations by about 2030. These two curves suggest that the world energy demand will level out at just over eight times the present figure. By the year 2000 the energy requirement would lie between 18–22 TW, about 2½ times the present figure.

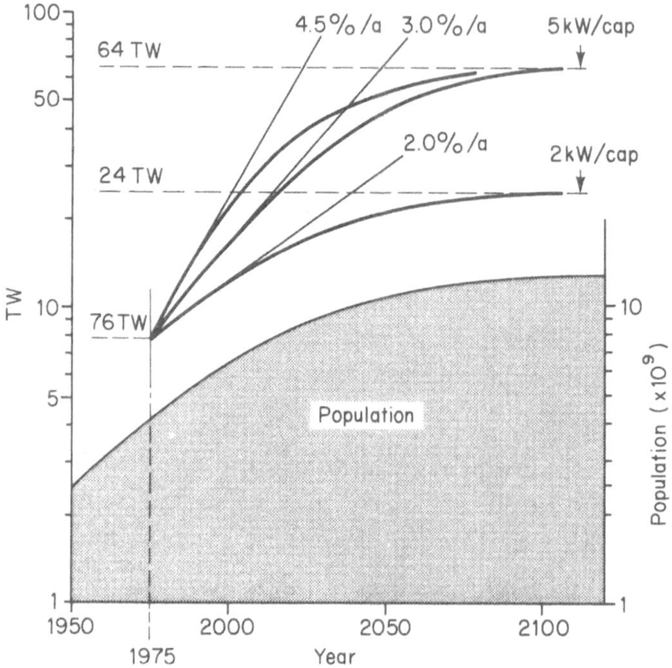

Figure 12: *Global energy scenarios. (Source: IIASA)*

A second illustration, this time based on varying assumptions for economic growth with their associated energy requirement, has been put forward in France by Robert Lattes, a member of the Club of Rome.

In Figure 13 the continuous line represents the most probable growth of GNP;[2] a historic figure of 5.2% per year from 1950 to 1975, 4.4% from 1975 to 2000 and 3% from 2000 to 2025. These growth rates are the lower figures for each of the alternative pathways. The upper figures are the associated energy growth rates, taken as 0.8% of economic growth over the period 1975 to 2000 and 0.6% from 2000 to 2025. Such a ratio of energy to GNP falling from the historic value of 1 over the period 1950–1975 represents a very considerable effort in improving the efficiency

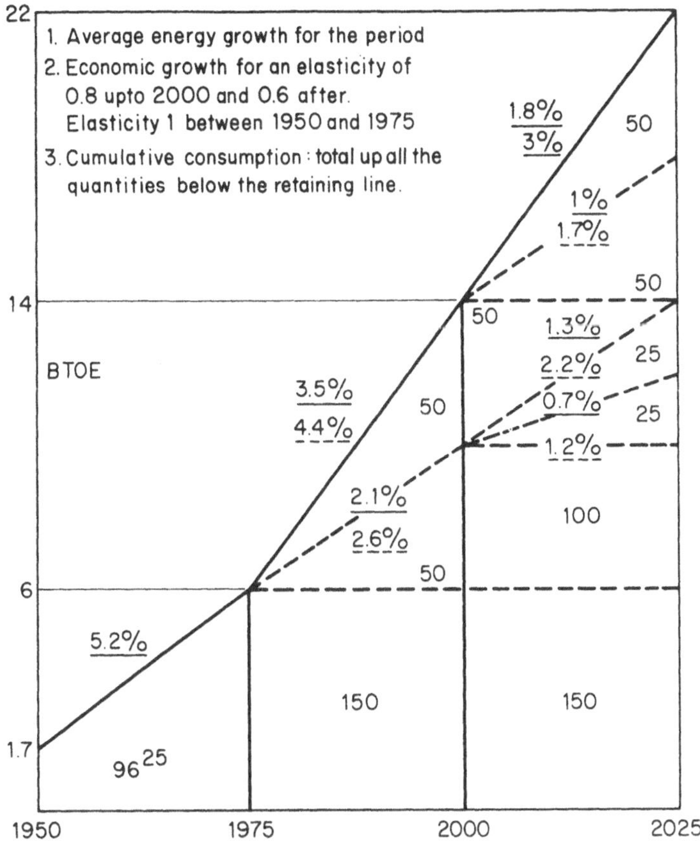

Figure 13: *Future annual world energy requirements (Source: Ref. 2)*

of energy use by conservation and other measures. The result of these projections is that for the most probable growth rate, the world energy demand would increase from 6,000 million tonnes oil equivalent (MTOE) in 1975 to 14,000 MTOE by 2000 – a factor of 2.3 – and 22,000 MTOE by the year 2025 – a factor of 3.7. For lower rates of growth the projections range from 10,000 up to 20,000 MTOE. The figures in the areas under each of the different pathways indicate the cumulative energy consumption during the period. These are (in billion TOE):

1950–1975	96.25
1975–2000	200–250
2000–2025	250–450

A third assessment of world energy demand comes from studies published by the World Energy Conference (WEC) in their 1978 report *World Energy: Looking Ahead to 2020.*[2] This reviews and comments upon a study carried out for the WEC by the Energy Research Group at the Cavendish Laboratory, Cambridge University which put forward a number of scenarios for the growth of energy demand. These have been published by the WEC together with summaries of reports on energy resources and energy conservation, in a separate report *World Energy Resources 1985–2020.*[4]

The Cavendish model starts from alternative assumptions for the annual growth rate of Gross World Product over the period 1975–2020 of 3.0% (low growth) and 4.1% (high growth) (Table 9).

Table 9: *Alternative assumptions for gross world product (GWP) growth rates in Cavendish report, %*

	OECD nations	*Centrally planned economy nations*	*Developing nations*	*World*
High growth	3.7	4.5	5.3	4.1
Low growth	2.8	3.2	3.8	3.0

Source: Ref. 3.

These models of high (H) and low (L) growth are further elaborated to take account of price response. H1 and L1 assume no price response, H2, H3 and L2, L3 a medium and high response to an increasing energy price where it is believed the average real price of energy to the final consumer will double over the next 25 years. This price response will be manifest by increasing conservation measures as well as a reduction in use. The H4 and L4 models assume that energy supplies will be restricted by limitations on the supply of oil, which will require some substitution of oil by other energy resources, while oil itself is reserved for premium use in transport and as a chemical raw material. H5 includes not only a high price response but also constraints on the use of oil and conservation measures that go beyond the normal consumer response to high energy prices. Such extreme conservation measures imply a significant change in the energy/GNP relation and will necessarily take some time before they can be fully introduced and become effective, H5 can then be seen (Figure 14) to show an appreciable reduction in energy demand by the year 2020. The results are illustrated in Figures 14[4] and 15.[3]

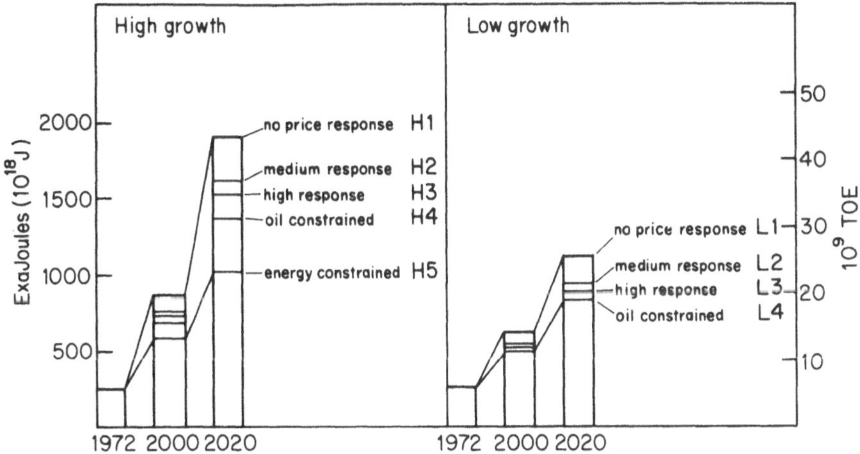

Figure 14: *World primary energy demand projections (Source: Ref. 4)*

The Conservation Commission of the WEC considered that the Cavendish projections did not take sufficient account of a changing relationship between economic growth and energy input and may then overestimate the energy demand for the industrialised countries and underestimate the energy demand of the developing countries. Taking rather high economic growth rates of 4.2% per year for OECD, 3.3% per year for the centrally planned countries and 5.7% per year for the developing countries the "Alternative Demand" projection was then put as in Table 10[3] and Figure 16.[3]

Table 10: *Alternative scenario demand projections*

	Primary energy demand, exajoules			
	OECD	*Centrally planned economies*	*Developing nations*	*World*
1972	150	66	27	243
1980	178	86	46	310
1990	212	120	86	418
2000	242	167	152	561
2010	262	233	253	748
2020	278	325	397	1000

Note: Only commercial energy is shown in the projections. 1 EJ = 10^{18} J

By the year 2000 the various figures, expressed in billions of

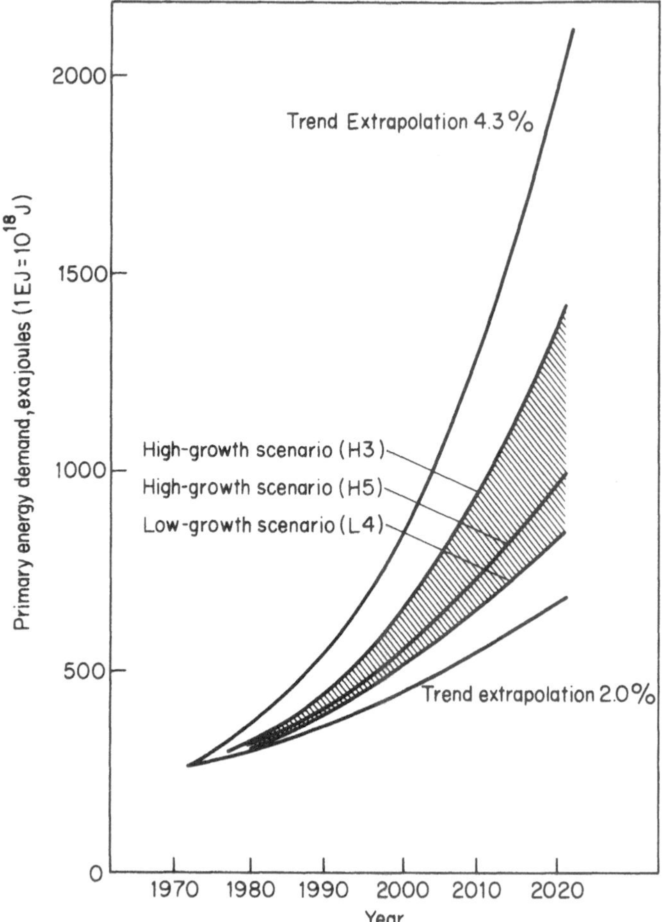

Figure 15: *Primary energy demand according to several projections. (Source: Ref. 3)*

tons of oil equivalent can be expressed as:

Cavendish	L4	12.0
	H1	20.0
	H2	17.5
Conservation Commission		12.7

The H1 figure which assumes that there is no response in energy demand to higher prices can be disregarded as unrealistic. Although it is emphasised by the WEC that these figures are not forecasts, for the purpose of comparison with other studies a mean figure of 15,000 MTOE is taken for the year 2000, approximately 2½ times the 1972 consumption. For the year 2020 the

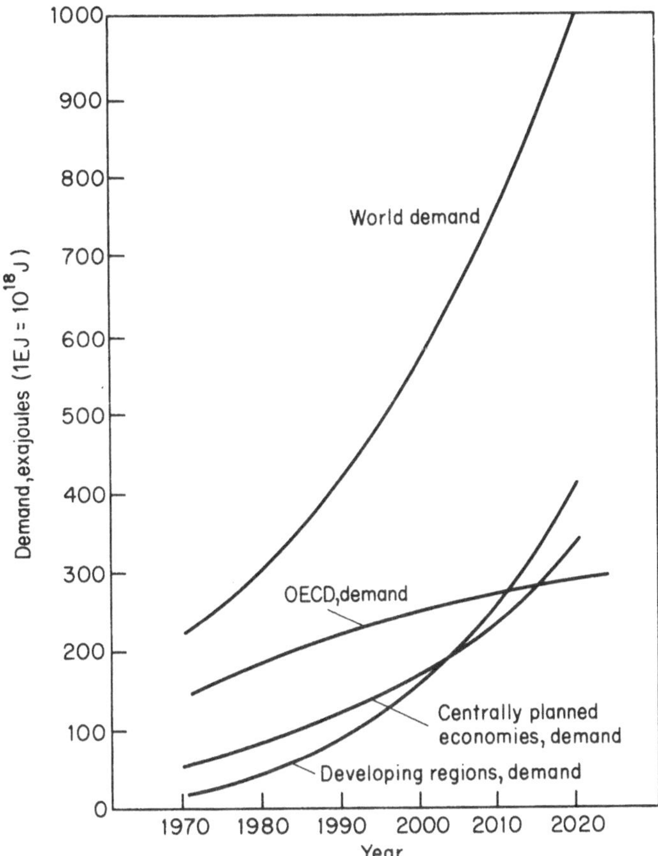

Figure 16: *The Conservation Commission's alternative demand projection (Source: Ref. 3)*

range of figures is from 18 to 36 billion TOE, 3 to 6 times the present level.

Another recent study *Energy – Global Prospects 1985–2000* published in 1977 is a report by the Workshop on Alternative Energy Strategies (WAES). This extensive analysis of world energy prospects to the year 2000 was carried out as an international project sponsored by the Massachusetts Institute of Technology in which some 75 individual experts from fifteen different countries took part. The WAES projections extend only to the World Outside Communist Areas (WOCA). Although the USSR and China are major energy producers, and suppliers, their trade in fuels to non-communist countries is small.

As in other projections one of the key factors is the rate of economic growth assumed over the period. WAES takes high and low figures. The high figure is 5.2%/year from 1977 to 1985 – a continuation of the economic growth and flourishing international trade of the 1960s – followed by a fall to 4%/year 1985–2000, mainly on the assumption that the developed countries would show declining population growth rates over this period. The low figure takes the initial rate to 1985 as 3.4%/year, only sufficiently higher than the estimated population growth of 2.2%/year to allow for a modest advance in GNP per capita. Any lower growth than this, it was suggested, could mean some periods of stagnation or recession and some measure of political and social instability especially in the "Third World" where hopes for material advancement are closely linked to world trade levels and to economic activity in the industrial countries. For 1985–2000 the low growth figure was reduced to 2.8%. Figure 17 in addition shows a breakdown into regional growth rates, with Japan, the OPEC countries and others having growth rates appreciably higher than N. America and W. Europe.[5]

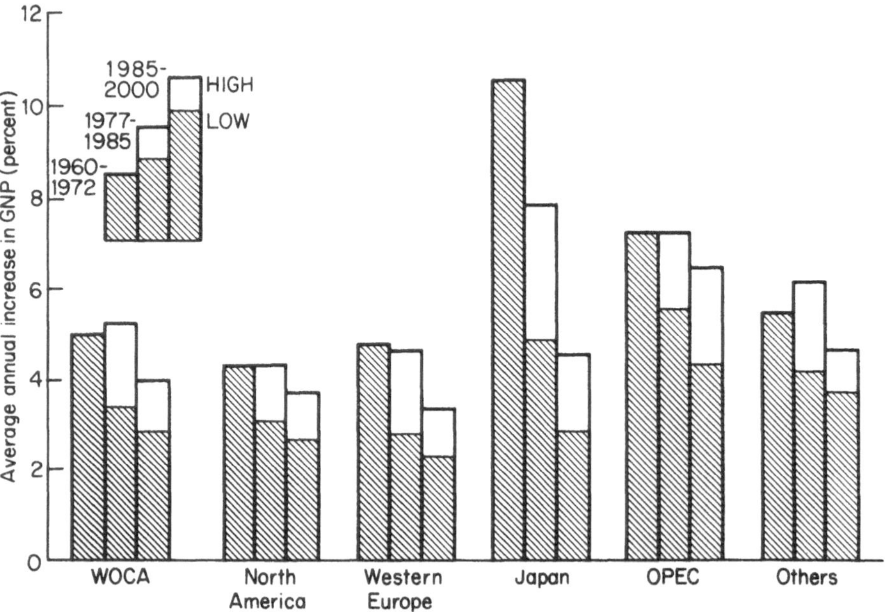

Figure 17: *Regional economic growth rates. (Source: Ref. 5)*

On this basis, and taking into account other variables which include energy price, government policies – e.g. to promote conservation – the WAES study estimates that the WOCA energy demand will rise from 4 MTOE in 1972 to between 8–10.4 MTOE by the year 2000, i.e. by a factor of 2–2.5 over 1972 levels.

If, to compare this with the other estimates, it is assumed that the demand in the communist countries (USSR and East Europe, China and Asian communist countries) will be 40% of the WOCA demand in the year 2000, the world figures can then be put at between 11 and 14.5 billion TOE.

In 1979 the union of the industrial federations of the EEC countries UNICE published a paper on expected energy demand for the Western world up to the year 2000 (Table 11).[6] For the

Table 11: *Predicted world energy demand (excluding communist countries), UNICE*

Energy source	*Actual*	*Estimated*	*High growth scenario*		*Low growth scenario*	
	1974	*1980*	*1990*	*2000*	*1990*	*2000*
Oil	47.1	50	66	72	67	68
Coal	16.0	18	22	27	20	22
Natural gas	16.4	19	23	27	21	24
Hydraulic, geothermal and solar	6.1	7	10	21	8	10
Nuclear	1.3	6	24	43*	20	26†
Total	86.9	100	145	190	136	150

All figures are in billions of barrels of oil equivalent per day. For nuclear, conversion is on the basis of specific consumption of thermal power stations. Economic growth rate hypothesis: 1980–1990 high growth 4.5 per cent, low growth 2.5 per cent; 1990–2000, high growth 3.5 per cent, low growth 2.0 per cent. *Represents installed power of 1450 GWe. †Represents installed power of 875 GWe.

high and low growth scenarios assumed, the figures range from 9.5 to 7.5 billion TOE – slightly below the WAES estimates given above. If these are adjusted in the same way the world figures would be 13.4 and 10.6 billion TOE.

The *Steam Coal Study* by the International Energy Agency published in 1978 is based upon an energy demand and supply estimate by OECD. This represents a forecast for the industrialised, developed countries of the world, OECD Europe, plus United States, Canada, Japan, Australia and New Zealand. Starting from 1976 the energy demand by the year 2000 increases by a factor of 1.9, despite assumptions of considerable advances in energy conservation and the efficiency of energy use. The Energy/GDP elasticity* figures assumed are shown in Table 12.[7]

*The annual per cent change in consumption of total primary energy for one per cent annual change in GDP.

Table 12: *Energy/GDP elasticities*

	1976–1985	*1985–1990*	*1990–2000*
OECD total	0.86	0.83	0.70
Seven largest countries	0.83	0.77	0.63

The total OECD demand figures are then (in MTOE):

1976	1985	1990	2000
3,613	4,864	5,592	6,913

Table 13: *Energy supply in the year 2000 (MTOE)*

	Coal	*Oil*	*Natural gas*	*Hydro-electricity*	*Nuclear*	*Others*	*Total*
Interfutures: High growth scenario							
OECD Europe	250	1,045	420	100	625	100	2,540
North America	750	1,205	450	175	650	100	3,300
Japan and Pacific	130	670	120	40	170	30	1,160
Total OECD	**1,130**	**2,920**	**990**	**315**	**1,415**	**230**	**7,000**
USSR	580	755	640	95	185	45	2,300
Eastern Europe	360	240	130	15	40	15	800
Total developed countries	**2,070**	**3,915**	**1,760**	**425**	**1,640**	**290**	**10,100**
China	950	555	105	50	75	45	1,780
OPEC and other LDCs	270	1,490	440	300	140	80	2,720
Total developing countries	**1,220**	**2,045**	**545**	**350**	**215**	**125**	**4,500**
Total world	**3,290**	**5,960**	**2,305**	**775**	**1,855**	**415**	**14,600**
Shares of the various energies	*22%*	*41%*	*16%*	*5%*	*13%*	*3%*	*100%*
IIASA: High growth scenario							
Total world	3,610	5,230	2,150	420	2,640	350	14,400
Shares of the various energies	*25%*	*36%*	*15%*	*3%*	*18%*	*3%*	*100%*
IIASA: Low growth scenario							
Total world	2,915	4,025	1,610	410	2,100	290	11,350
Shares of the various energies	*26%*	*35%*	*14%*	*4%*	*18%*	*3%*	*100%*

Source: Ref. 8.
The figures refer to the high growth scenarios of interfutures and to the high and low growth scenarios of the International Institute for Applied Systems (IIASA). In the Interfutures scenario of traditional moderate growth not included here, world consumption of energy is about 13,000 MTOE.

Another OECD project is Interfutures,[8] a 3-year study of "the future development of advanced industrialised societies in harmony with that of the developing countries". This study, completed in 1979, was carried out by a 15-man research team drawn from 10 countries, supported by consultants, an advisory panel representing economics, business, sociology and other disciplines, and with the whole project under the guidance of a governmental steering committee.

The conclusions on energy supply for the year 2000 are given in Table 13 which also includes high and low figures from an IIASA study.[8]

Table 14: *Summary of energy supply in the year 2000*

	BTOE Year 2000 world consumption	*Increase over 1975*
IIASA		
Increase from 2 to 5 kW per capita		
fast transition 4.5%/year	15.5	2.5
slow transition 3%/year	12.5	2.0
Lattes		
high growth	14.0	2.2
low growth	10.0	1.6
WEC		
Cavendish		
L4	12.0	1.9
H2	17.5	2.8
WEC		
Conservation Commission	12.7	2.0
Interfutures		
high growth	14.6	2.3
low growth	13.0	2.1
IIASA, 1978		
high growth	14.4	2.3
low growth	11.35	1.8
WAES		
WOCA only, low 8 adjusted	11.2	1.8
high 10.4 adjusted	14.6	2.3
UNICE		
WOCA only, low 7.5 adjusted	10.5	1.7
high 9.5 adjusted	13.3	2.1
IEA		
Steam Coal Study, OECD 6.9 adjusted (x2)	14.0	2.2

While the energy consumption of the OECD area is put at almost double the 1976 level, that of the developing countries is expected to increase by a factor of between 5–7.

These figures can then be summarised (Table 14). For comparison they are all expressed in the same unit, billions (10^9) of tonnes of oil equivalent (BTOE). To show the predicted increase they are compared with a base figure of 6.3 BTOE for the 1975 world consumption, and 4.4 BTOE for WOCA. To adjust the year 2000 WOCA figures to a world level the centrally planned countries are assumed to have an energy consumption at that time of 40% of WOCA. From this it can be seen that there is a striking consistency about all these figures which indicate that the expected world energy demand in 2000 will be about twice the 1975 figure.

HOW CAN THE DEMAND BE MET?

It should be emphasised that the estimates of future demand are not forecasts or predictions of what will happen, they are more in the nature of assessments or projections of how much energy will be required to meet a modest but continuing increase of economic welfare in a world with an expanding population. A prediction is a precise statement of what will happen; a projection is an open-ended possibility derived from a careful assessment of past experience. On the other hand a projection, if it is used to guide or determine policy will tend – where this is possible – to be self-fulfilling. As has been pointed out energy demand is largely determined by the availability of supply. It is the problem of the chicken and the egg. Estimates of demand will be conditioned by what it is believed will be available. The figures given can then be considered as providing targets to be aimed for to ensure economic stability. To see where this energy can come from, the contributions to be made from the different possible sources is examined.

For the World Energy Conference Conservation Commission the figures considered are not those of the Alternative Demand Scenario, but the higher figures given as the potential world primary energy production (Table 15).

It is noted that the production rates shown will "require a vigorous development of every primary energy resource" and it is emphasised that this will be "a most challenging task, requiring the mobilisation of capital, labour, materials and technical and managerial skills on a scale never before undertaken".

These figures can then be taken as representing an upper limit

Table 15: *Potential world primary energy production (exajoules)*

Resource	*1972*	*1985*	*2000*	*2020*
Coal	66	115	170	259
Oil	115	216	195	106
Gas	46	77	143	125
Nuclear	2	23	88	314
Hydraulic	14	24	34	56
Unconventional oil and gas	0	0	4	40
Renewable, solar, geothermal, biomass	26	33	56	100
Total	269	488	690	1000

of energy supply that might be achieved provided a sufficient effort is put in hand without delay.

For each energy source the figures to the year 2000 can be expressed as a percentage of the total and as an increase over the 1972 values (Table 16).

Table 16: *Energy sources as a percentage of total in 2000 and increase over 1972*

	% share in year 2000	*Increase over 1972*
Coal	24.6	x 2.6
Oil	28.3	x 1.7
Gas	20.7	x 3.1
Nuclear	12.8	x 44
Hydro + other	13.6	x 2.4

The UNICE figures (for the WOCA only) are given in Table 17. These again can be expressed in terms of market share and increase, this time over 1974 consumption, taking the high growth figure.

Table 17: *UNICE figures as a percentage of total in 2000 and increase over 1974*

	% share in year 2000	*Increase over 1974 consumption*
Coal	14.2	x 1.7
Oil	37.9	x 1.5
Gas	14.2	x 1.6
Nuclear	22.6	x 33
Hydro + other	11.1	x 3.4

The world figures given in the OECD Interfutures Study and the comparison given there with the 1978 IIASA estimates are shown in Table 18.

Table 18: *Interfutures and IIASA figures as a percentage of total in 2000 and increase over 1972*

	% share in 2000		
	Interfutures	*IIASA High*	*IIASA Low*
Coal	22	25	26
Oil	41	36	35
Gas	16	15	14
Nuclear	13	18	18
Hydro + other	8	6	7
	Increase over 1972 base		
Coal	x 2.2	x 2.4	x 1.9
Oil	x 2.3	x 2.0	x 1.5
Gas	x 2.2	x 2.0	x 1.5
Nuclear	x 41	x 58	x 46
Hydro + other	x 1.3	x 1.0	x 1.0

For the IEA Steam Coal Study based on OECD estimates, the figures given for indigenous supply and imports are combined, and mean figures taken where ranges are given according to high or low nuclear supply. The supply pattern for the year 2000 results is then as in Table 19.

Table 19: *Supply pattern for combined OECD indigenous supply and import figures*

	% share	*Actual increase over 1976 production*
Coal	20.5	x 1.9
Oil + Natural gas	57.5	x 1.52
Nuclear	15.15	x 11.9
Hydro + other	7.4	x 2.2
	100.0	x 1.9

These figures assume that the OECD import requirements for 2000 are met in full despite the fact that this would then leave the rest of the world with a deficit of some 1,220 million tons of oil. But it is noted that in reality a gap between supply and demand cannot occur in the world market. The potential excess demand is meant to display the magnitude of unsatisfied energy demand that must be met by an extraordinary effort in reducing demand and expanding supply.

These figures can be summarised as in Tables 20 and 21.

Table 20: *Summary of source energy as a percentage of total in 2000*

	% share of estimated total energy by year 2000				
	Coal	*Oil*	*Natural gas*	*Nuclear*	*Hydro + other*
WEC	24.6	28.3	20.7	12.7	13.6
UNICE (WOCA only)	14.2	37.8	14.2	22.6	11.2
WAES 1 (WOCA only)	17.0	46.8	14.0	14.2	7.9
WAES 2	12.8	44.2	13.0	23.5	6.5
OECD (OECD only)	20.5	45.0	12.5	15.2	7.2
Interfutures	22.0	41.0	16.0	13.0	8.0
IIASA (1978), high	25.0	36.0	15.0	18.0	6.0
IIASA (1978), low	26.0	35.0	14.0	18.0	7.0

Table 21: *Source energy in 2000 as an increase over base figures*

	Base date	*Increase over actual figures*				
		Coal	*Oil*	*Natural gas*	*Nuclear*	*Hydro + other*
WEC	1972	x 2.58	x 1.7	x 3.1	x 44.0	x 2.35
UNICE	1974	1.70	1.5	1.6	33.0	3.50
WAES 1	1972	2.30	2.1	1.8	40.0	2.70
WAES 2	1972	1.50	1.7	1.4	55.0	2.00
OECD	1976	1.90	2.05	1.26	11.9[a]	2.20
Interfutures	1972	2.20	2.3	2.2	41.0	1.30
IIASA (1978), high	1972	2.40	2.0	2.0	58.0	1.00
IIASA (1978), low	1972	1.90	1.5	1.5	46.0	1.00

[a] This low figure is a consequence of the rapid expansion of nuclear capacity between 1972 and 1976, the base date for the OECD figures.

These figures show a remarkable consistency, given the uncertainties involved. The average expansion assumed is then: Coal x 2.1; Oil x 1.85; Gas x 1.85; Nuclear x 45 (omitting the OECD figure) and Hydro + renewables x 2.1. But there must now be considerable doubt as to whether such expansion figures can be achieved by the year 2000. The output of oil and gas is likely to be restricted for political and policy reasons and may never exceed present levels: it takes time to build up a substantial international trade in coal and to bring new mines into production: there are delays in developing a synthetic oil from coal industry: nuclear power expansion has been slowed down.

The question of delay has been considered by the World Energy Conference. The maximum rates of expansion of production were assumed to have started from the year 1972 in response to the perceived need. The effect of a 10-year delay is shown in Table 22. But of this 10 years we have already lost seven, and since the WEC point out they have already used maximum production rates, there is no way in which the time lost in delayed response can be made up for by enhanced production. "We are now more short of time than of energy."[3]

Table 22: *Energy balances*

	Supply: global production of primary energy, exajoules/year	
	Prompt response, $t_D = 0$	*Delayed response,* $t_D = 10$
1972	242	242
1980	361	341
1990	491	440
2000	634	550
2010	797	659
2020	1004	814

REFERENCES

1. Second Status Report on the IIASA project on energy systems 1975, W. Häfele *et al.*, International Institute for Applied Systems Analysis, 2361, Laxenburg, Austria.
2. Thierry de Montbrial, *L'Energie: Le Compte a Rebours*, J.-C. Lattes, Paris.
3. *World Energy: Looking ahead to 2020*, IPC Science and Technology Press, Guildford.
4. *World Energy Resources*, IPC Science and Technology Press, Guildford.
5. *Energy: Global Prospects 1985–2000*, Workshop on Alternative Energy Strategies, McGraw-hill, New York.
6. UNICE Brussels, reported in Nuclear Engineering International (May 1979).
7. *Steam Coal Prospects to 2000*, OECD, 2 rue Andre-Pascal, 75775 Paris.
8. *OECD Observer*, No. 100 (September 1979).

3
The availability of oil

The 20th century, the age of oil, has seen the rise and will see the fall of oil as the principal source of world energy.

Even though the eventual depletion of world oil has been previously predicted, little heed has hitherto been taken of these warnings, as past forecasts of imminent scarcity have always been forestalled by new discoveries or the extension of existing reserves. And particularly in the industrial countries, the main energy consumers, there has been a marked acceleration in the use of oil since 1960, with a swing away from coal. Coal which supplied nearly 50% of world energy in 1960 was by 1971 displaced by oil, the source of 40%, and by natural gas, 20%, of a much larger total demand (Table 23).

Table 23: *Development of the pattern of primary energy consumption*

	EUR-9		*USSR*		*USA*		*JAPAN*		*WORLD*	
	1960	*1971*	*1960*	*1971*	*1960*	*1971*	*1960*	*1971*	*1960*	*1971*
	millions of tons of petroleum equivalent									
Hard coal and lignite	327	233	269	309	251	316	44	56	1,543	1,671
Crude petroleum	146	489	110	257	415	679	27	172	926	2,120
Natural gas	9	80	42	201	338	607	1	4	434	1,059
Primary electrical energy	30	37	16	30	49	76	19	22	218	328
TOTAL	512	840	437	798	1,053	1,679	90	254	3,121	5,178
	expressed as a %									
Hard coal and lignite	64.0	27.7	61.6	38.7	23.9	18.8	48.6	22.0	49.4	32.3
Crude petroleum	28.5	58.2	25.1	32.2	39.4	40.5	29.7	67.5	29.7	40.9
Natural gas	1.7	9.5	9.6	25.2	32.0	36.1	1.0	1.5	13.9	20.5
Primary electrical energy	5.8	4.4	3.7	3.8	4.6	4.5	20.9	8.8	7.0	6.3
TOTAL	100.0	100.0	100.0	100.0	100.0	100.0	100.0	100.0	100.0	100.0

Source: Ref. 1.

The expansion of world energy consumption over the period from 1960 has been almost entirely met by increases in the supply of oil and natural gas.

In Europe over this ten year period the consumption of coal has decreased by more than 2% per year, while the consumption of petroleum has risen at an average rate of 14%.

The rise in primary energy consumption and the swing to oil has meant that Europe, the USA and Japan have had to resort increasingly to obtaining supplies from outside sources, so that in ten years 1960–1971 the net energy imports of the EEC and the USA more than trebled, while those of Japan increased sevenfold (Table 24).

Table 24: *Dependence on imported energy*

	Net imports less bunkers (MTOE)		*Degree of dependence on imported energy*[a]	
	1960	1971	1960	1971
EUR-6	98	380	29.9	62.9
EUR-9	158	509	30.9	60.3
USA	48	157	4.6	9.3
Japan	31	215	34.7	84.7

[a]Defined by the ratio = $\frac{\text{net imports less bunkers}}{\text{primary energy consumption}}$ as a %.

Source: Ref. 1.

The only exception to this general tendency among the industrialised countries is the USSR which has an export balance of about 100 million tons of petroleum equivalent. But these exports go mainly to the CMEA countries. Little is available for the rest of the world.

The present dependence on imported fuels for the OECD countries is given in Table 25.

Whilst these imports could be tolerated, or even encouraged, when the oil price was low, this is no longer the case. The age of cheap energy was brought to a close by the events of October 1973, the Yom Kippur War. Yet this first warning has been ignored, and after the initial shock of the oil price rise was overcome and the effects of the recession absorbed there was until the end of 1978 a transition to apparent normality; a calming of the fears raised in 1973 was fostered by the short term surplus of petroleum which developed during the years up to 1978. But this surplus was in part a consequence of the oil price induced recession which has led to a reduction in industrial output and would no doubt have disappeared within a year or so of an industrial recovery. This process was however interrupted by the 1978 revolution in Iran

Table 25: *Energy balance (MTOE) 1975*

	Total energy requirements	*Indigenous production*	*Self sufficiency, %*
USA	1,690.15	1,455.96	86.0
OECD Europe	1,116.26	491.96	44.0
Austria	22.86	10.30	45.0
Belgium	41.83	6.91	16.5
Luxembourg	3.99	0.16	4.0
Denmark	17.70	0.15	0.8
Finland	22.30	6.63	30.0
France	168.06	43.29	26.0
Germany	243.49	118.92	49.0
Ireland	7.08	1.55	22.0
Italy	127.03	25.23	20.0
Netherlands	58.99	72.29	123.0
Norway	19.15	21.22	111.0
Portugal	8.20	1.65	20.0
Spain	60.68	17.29	29.0
Sweden	49.10	19.37	40.0
Switzerland	22.44	9.68	43.0
UK [a]	203.22	116.95	58.0

[a]With the development of North Sea oil, the UK is expected to reach self-sufficiency for some years from 1980 onwards and even be a net exporter of energy.

which has cast doubt over the availability of some 10% of the world's supply of oil and led at once – at least in the spot market – to oil prices rising steeply, once again reawakening the sense of urgency and fears over world energy supply problems. Supplies from Iran are now restarting, but at a lower level and with some uncertainty over future production, but the general price rise decided by OPEC with additional surcharges being imposed by most of the producing countries continues. Even so there is still a reluctance by the oil importing countries to face the unpleasant reality. The price increases so far are tolerable and indeed up to mid-1979 have done little more than keep pace with the decline in the value of the dollar due to inflation (see Figure 18).[2] Reductions in output have been cushioned by the stocks which the Governments of most industrial countries determined should be built up after the 1973 interruption of supplies to cover such contingency as the Iranian revolution. But the realisation that the exhaustion of major oil producing areas is in sight is now inescapable, and while the industrial countries will try to avoid this by bidding competitively for their supplies the fact will be brought home by the producing countries imposing restrictions on output.*

* OPEC plan for production cuts to maintain prices, *Financial Times*, 21 March 1980; Libya, Iraq, Kuwait and Nigeria cut oil production, *Financial Times*, 22 March 1980.

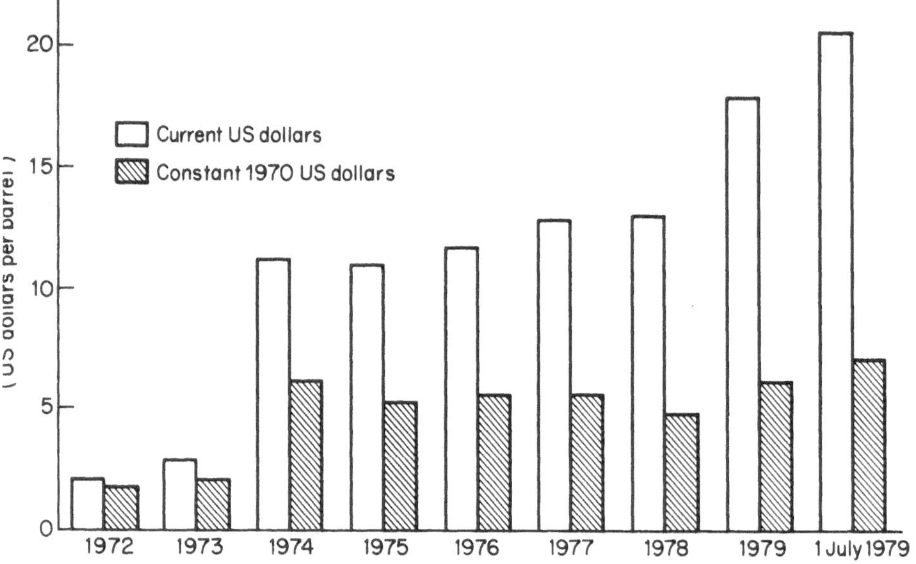

Figure 18: *Petroleum prices, 1972–1979. (Source: Ref. 2)*

Any need the producers have for increased oil revenues will be met by increasing the price rather than the production. It is the spokesmen for the Arab oil producers and for OPEC who are urging that the industrial countries should reduce their consumption of, and dependence on, oil.

As well as providing the fuel for the rest of the world, the governments of the oil producing countries also have a responsibility with respect to the economic and social development of their own countries. Even if the oil producers have now accumulated gross foreign assets approaching 200 billion dollars, most of them are still developing countries, struggling to establish an economic and social infrastructure so as to be able to provide such benefits as education, medical care, transport systems. They are also seeking to establish alternative means of earning foreign exchange so that they will be less dependent on a single, depleting resource. For the Arab countries for instance about 90% of their income comes from exports of crude oil, oil products and natural gas. Although this income may increase in the short term as the world energy market tightens and prices rise, crude oil is a declining asset. 20–30 years of active lifespan for the proven oil reserves does not give much time for the Arab world to develop their economies so as to cushion the blow of declining oil revenues. Much can, of course, be done by a wise programme of investment of oil funds in the manufacturing industry of the West, but there could be a growing resistance in some countries to fears that an excessive proportion of their industrial assets is being acquired by foreign owners. There is also a desire on the part of the oil producing governments to provide employment for their people.

There is then a considerable industrial development now starting the Arab countries.* Projects already established or planned will produce over 1 million tonnes of primary aluminium products. Using imported bauxite and cheap flare gas over 120,000 tonnes/year of primary aluminium are already being produced in Bahrain. Egypt too also has an output of some 50,000 tonnes/year. New aluminium production capacity is now being planned for Saudi Arabia 200,000 tonnes, Iraq 150,000 tonnes, Dubai 130,000 tonnes and Abu Dhabi 350,000 tonnes.

For steel, presently operating and planned projects will give a production of well over 3 million tonnes/year by the 1980s. In Iraq, for instance, the State Company for Iron and Steel is building a large steel complex at Khor Al-Zobair with two direct reduction plants with a combined capacity of more than 1 million tonnes/year, which will use liquid petroleum gas from the Rumaila Oilfields. In Saudi Arabia there are plans for the construction, at a cost of 400 million dollars, of a plant to produce 900,000 tonnes/year of steel billets. Elsewhere in the Arab world there are similar plans – though not always of such large output – in Jordan, Kuwait, the Lebanon, Oman, Egypt, Qatar, Dubai and Abu Dhabi.

The Arab States are also investing over 50 billion dollars in oil refining and petrochemical production. By the early 1980s Kuwait, Algeria, Egypt and Libya will be producing over a million tonnes/year of plastic and petrochemical products. Iraq may soon become the leading Arab producer of nitrogen and phosphate-based fertilisers with an annual output of over 7 million tonnes.

If these projects develop as planned over the next ten years or so the oil producing countries will themselves become major consumers of energy, rather than merely energy exporters.

At the beginning of the 1970s the developing countries, including members of OPEC consumed some 15% of the WOCA (World Outside Communist Area) total energy; by the year 2000 industrialisation and development will push their share up to 25%.

* Shakirullah Durani, "The Use of Arab Funds to Finance Energy Projects", Financing the World's Energy Industries to the Year 2000, Graham & Trotman Conference, London, 1979.

The goal of the OPEC countries is to create an industrial society which will enable them to maintain a favourable position in the world economic and political system at a time when oil production will be declining. This will be a slow process, and the dangers of pushing it too fast are now evident from the experience of Iran, where the social stresses and strains arising from the attempt to transform Iran, perhaps too quickly, into a modern industrial state can be considered to be a contributory factor to the 1978 revolution. To reach the goal may then take longer than the expected life span of their oil reserves at present rates of production. It must then be expected that the OPEC countries will reduce production in order to ensure that their oil wells are not exhausted before the long range goal is achieved. It has been suggested that a reduction of output of 50–70% would be required to extend the life of the oil by some 50 to 100 years. But at the same time an increasing proportion of this oil will be consumed within the OPEC countries instead of being exported. The probability that this policy will be carried out is high; it could only be effectively opposed by the political or other action of one or more of the major powers. But both the United States and the USSR still have a substantial oil production and large coal reserves.

It is certain that more oil will be found and recovered; as the price rises a greater effort will be put into exploration for new deposits outside the main producing areas; a greater use will be made of enhanced recovery methods to increase the quantity that can be extracted from existing oil fields and some of the "unconventional" heavy oil requiring steam or heat injection will be brought into production; but the quantities are unlikely to have much impact on demand, the cost could be high, and the "life" short. North Sea oil for instance is expensive to produce, but although its availability will put the UK in a favourable position, it will only be over a limited period during the 1980s that North Sea production will exceed the domestic demand. By the early 1990s the UK will wish to resume oil imports at an increasing rate with a potential demand rising to 35–50 million tons/year by the end of the century – if supplies can be found at that time.

The major effects of any restriction of oil supplies will be felt in continental western Europe and Japan which will be faced with a supply deficit of roughly 50–70% of their present consumption, probably within the next ten years. This will come as a combination of price increase and supply shortage.[3] (See Figure 19.)

It is then rash to assume that at a time when the decline of their reserves will be starting to become evident, the oil producing

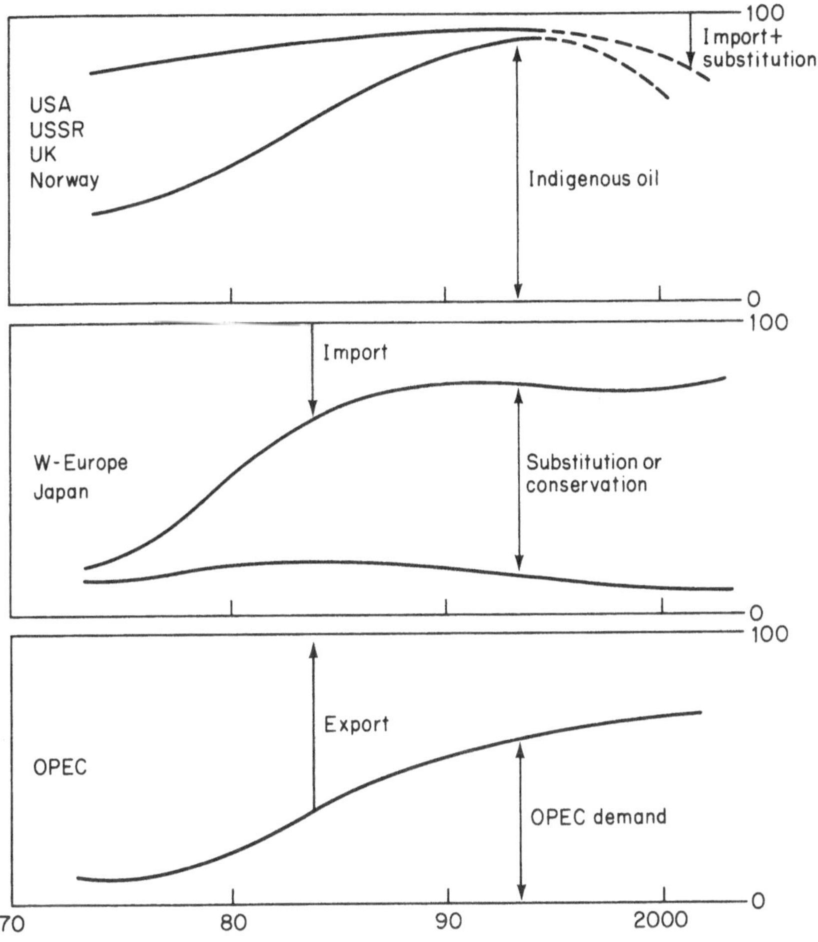

Figure 19: *Oil supply prognosis. (Source: Ref. 3)*

countries will continue to export oil to the industrial nations on the same scale as in the past. Indeed it may already be the case that the present world oil production of 23.65 billion barrels is approaching the upper limit.* This view is now supported by Professor Schneider, a member of the WAES study. In a paper to the 4th Annual Symposium of the Uranium Institute he said that prospects for expansion of the world supply of oil must be viewed with scepticism. This is not so much because of the ratio of current oil consumption to proven oil reserves, since with technical advances and increasing oil prices it can be assumed

* 1979 production figures in billion barrels (*Petroleum Economist*, March 1980): OPEC 11.2 (Saudi Arabia 3.5); Other free world 7.2 (USA 3.7); USSR 4.3; World 23.7.

that more oil could become available and any limit on production by reserves can be pushed further into the future. The doubt over further expansion of oil supplies lies in the expected reaction of the major oil producers "In important oil-rich countries, such as Iran and Saudi-Arabia, one can identify a definite reluctance to expand productive capacities" and he drew attention to a reduced investment in new oil production facilities in the OPEC countries and said that "the probability of an increase in oil output of 30–40% on today's figure, and a stabilisation at this level up to the turn of the century must be considered very unlikely" it was then "risky to build such an increase of oil production into energy planning".

The effect of this is shown in Figure 20 which illustrates the gap of 1–2 billion tons which will appear by the year 2000 between the WAES forecasts and a steady production at the 1978 level.

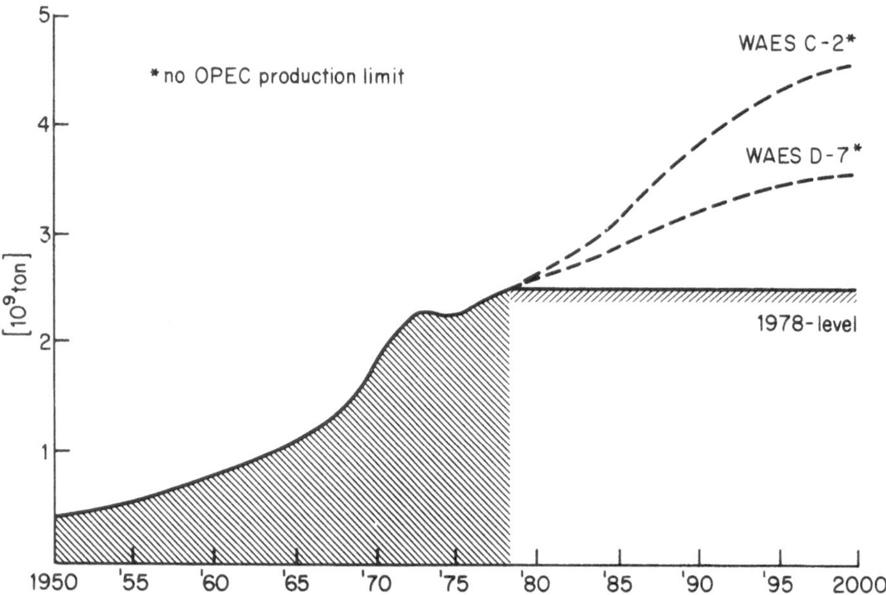

Figure 20: *Oil demand and availability (WOCA). (Source: Ref. 4)*

This may seem a pessimistic forecast, yet it is in accord with the view of a staff report to the US Senate Committee on Foreign Relations, on the future of Saudi Arabian Oil Production, which was published in April 1979. This concluded that:

> "the United States should not base its energy plans on the premise that Saudi Arabia, as residual supplier, will produce enough oil to supply the needs of the United States or the world economy over the next two decades at anticipated rates of oil consumption . . . it would be imprudent for the United States to plan on a change in Saudi Arabian oil development plans above 12 MBD".

At this level the operating company, Aramco, believes that production could be maintained for 15–20 years before irreversibly declining. But the actual production figures are even lower: the first quarter of 1979 was at the rate of 9.8 MBD; by December 1979 there was relief that Saudi production will continue at a "raised" level of 9.5 MBD, at least for the first quarter of 1980.

For natural gas, as for oil, the WAES expects demand to outrun supply. By the year 2000 this study predicts that natural gas production in the major consuming countries will probably decline to about two-thirds of the 1972 level. In North America production will reach a plateau and start to decline before 1985. Expanded production from the North Sea will take W. Europe up to the latter 1980s before the decline begins – unless reserves are much larger than now estimated. Demand is expected to increase and N. America, W. Europe and Japan together might require natural gas imports of about 3.3 MBDOE in 1985 (165 MTOE) and 8.4 MBDOE by the year 2000 if the demand is to be met – with supplies coming from the USSR and OPEC. The overall WOCA demand figure is shown in Figure 21 rising to 27 MBDOE or 1,350 MTOE/year by the year 2000 but with the WOCA production and planned imports falling short of this by some 4 MBD.[5]

The world figures as given in the World Energy Conference report suggest that total world supply and demand should be in reasonable balance up to the year 2020 (Table 26). But there are marked differences within the different regions which will require a sizeable international trade in natural gas.

But as with oil, natural gas production could be restricted for political reasons or to give priority to new industrial consumption projects in the producing countries. The refusal by Iran to continue with the partially constructed IGAT-2 gas pipe lines to the USSR is an indication of the uncertainties that will arise. In this particular case the further export of gas from the USSR to Europe which formed part of the overall arrangement will also be affected. With the OPEC countries holding some 55% of the proven natural gas reserves of WOCA it is probable that policies on the export of this gas will follow closely those of oil. Indeed it is reported (*Financial Times*, 20 September 1979) that the OPEC countries are trying to "unify" gas prices and to bring them

Table 26: *World gas demand and potential production from conventional resources (exajoules or $ft^3 \times 10^{12}$) ($1\ ft^3 = 0.028\ m^3$)*

	1985		2000		2020	
	Demand	*Production potential*	*Demand*	*Production potential*	*Demand*	*Production potential*
OECD						
N. America	22.0	29.7	27.0	27.3	29.0	10.7
W. Europe	6.0	9.6	11.0	8.7	13.0	2.2
JANZ[a]	0.8	0.4	3.5	2.1	5.5	4.6
Sub-total	28.8	39.7	41.5	38.1	47.5	17.5
CPE						
USSR/E. Europe	16.0	21.8	30.0	55.7	39.0	28.5
China/Asia	1.5	1.7	2.5	2.9	11.0	6.1
Sub-total	17.5	23.5	32.5	58.6	50.0	34.6
Developing nations						
OPEC	3.0	11.9	5.5	39.4	7.0	63.3
Non-OPEC	3.5	3.7	6.5	7.4	9.0	9.6
Sub-total	6.5	15.6	12.0	46.8	16.0	72.9
Total	52.8	78.8	86.0	143.5	113.5	125.0

[a] Japan, Australia, New Zealand

in line with those charged for petroleum products. If the commercial policies are unified then it must be assumed the political policies will be also.

Yet the demand and supply forecasts reviewed in Chapter 2 assume that oil and gas will meet about half the world's expected energy needs by the year 2000 (from 49% for the WEC to 61% in the WAES high case). This would require the present production to be doubled. This must now be considered most unlikely. The best that can be hoped for is that the amount of oil available on world markets will be held at present levels before it inevitably starts to decline. Oil production will from now on be regulated to meet the needs of the producers rather than as previously the needs of the customers. Shortages of oil are then likely to arise within the next few years, and this will come in the first place from restriction on increasing output imposed by those producers who are concerned over their own future needs and economic development, rather than from physical limits to production. These shortages will initially be felt most severely in the oil importing countries of continental Europe and Japan. But the recession that will be induced in those countries will affect the whole world.

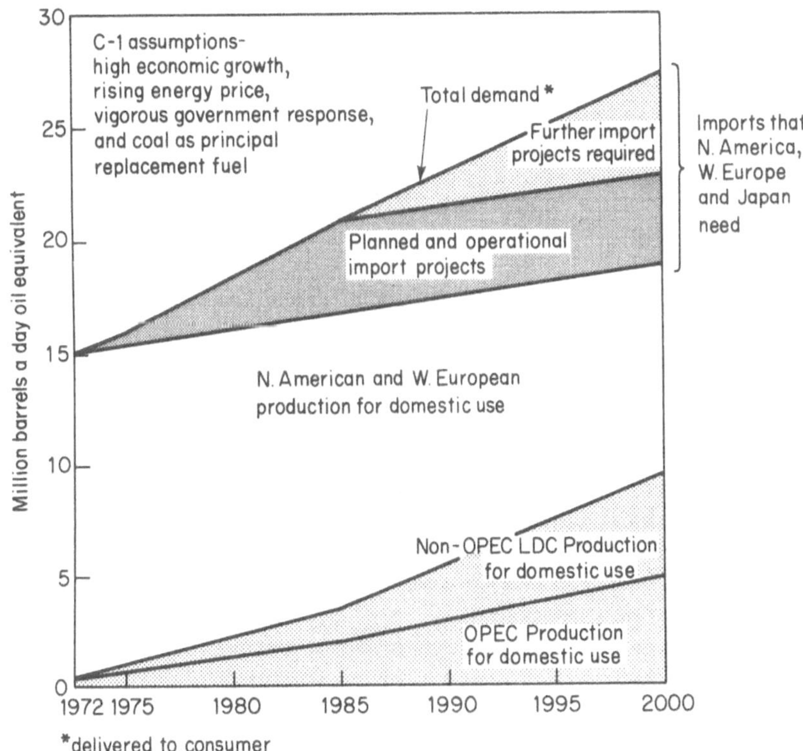

Figure 21: *Natural gas supply and demand (WOCA).*
** = delivered to consumer. (Source: Ref. 5)*

REFERENCES

1. Eurostat publications, Luxembourg.
2. *World Development Report*, World Bank, 1979.
3. U. La Roche, *Energy Scenario*, Brown Boveri, Baden, Switzerland.
4. H. Schneider, 4th Annual Symposium of the Uranium Institute, September 1979.
5. *Energy: Global Prospects 1985–2000*, Workshop on Alternative Energy Strategies, McGraw-Hill, New York.
6. *World Energy: Looking Ahead to 2020*, Conservation Commission of World Energy Conference, 1978.

4
Coal

There are vast reserves of coal in the world, far in excess of those of the other fossil fuels, and these provide a resource base sufficient to support coal extraction programmes well into the next century. And these reserves are widely distributed; the USA, USSR and China have the very largest deposits, but there are substantial coal reserves in Europe, Australia, Southern Africa, India and Canada.

The problems with coal relate more to its acceptance, use and international marketing rather than with the production. Consumer preference is for the cleaner, more easily handled oil fuel which does not require large coal handling facilities and ash disposal. Oil has always been more expensive than coal in the USA, and in spite of that it has progressively displaced coal in the energy market even in the most massive use of it for electricity production. Figure 22 shows how the market share of coal in the US has declined relative to oil and gas. The difficulties of reversing this long term trend should not be underestimated. But although coal now takes a declining share of the expanding energy market, it is by no means a dying industry, on the contrary production still continues to increase (Figure 23).

Over the period 1950–1975 the actual increase was at the rate of 2.3%/year to reach a world total figure of 2,593 million tons.

The major producing countries, and their future production as estimated by the World Energy Conference are given in Table 27.[1]

It can be noted that approximately half the total production of coal comes from Russia, China and other Eastern Bloc countries.

A detailed assessment of the potential expansion of coal use for the OECD countries is given in the International Energy Agency report *Steam Coal: Prospects to 2000.*[2] In the "reference" case this report sets out forecasts for the OECD Energy Demand and Supply using alternative high and low estimates of nuclear power production (Table 28).

In accordance with other forecasts this shows the OECD energy demand increasing nearly two-fold over the period 1976–2000.

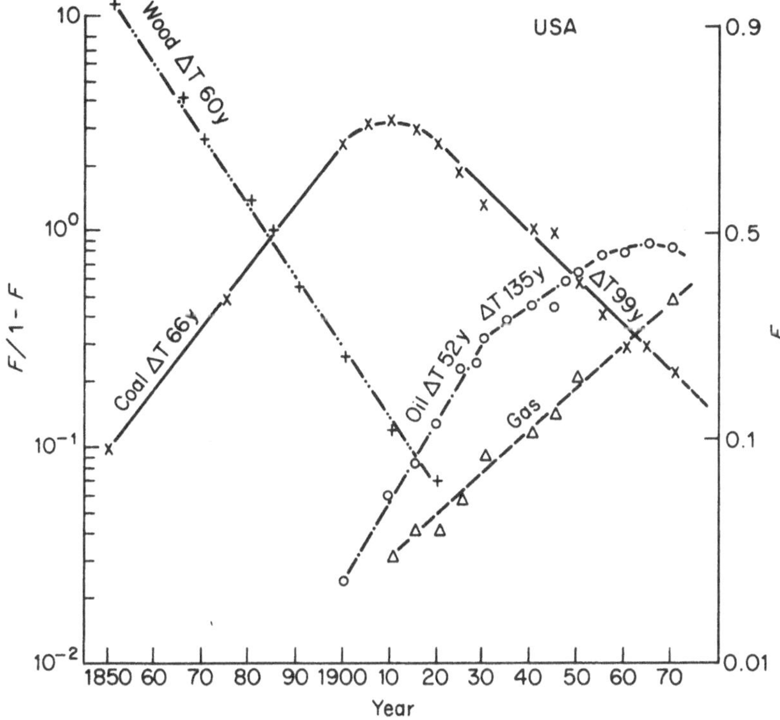

Figure 22: *The market share for primary energy consumption in the US.* F = *fraction of total market taken by each energy source. (Source: IIASA)*

Table 27: *Estimated future production by the main coal-producing countries for 1985, 2000 and 2020*

Country	*Coal production (MTCE)*			
	1975	*1985*	*2000*	*2020*
Australia	69	150	300	400
Canada	23	35	115	200
China (People's Republic)	349	725	1,200	1,800
Germany (Federal Republic)	126	129	145	155
Great Britain	129	137	173	200
India	73	135	235	500
Japan[a]	19	20	20	20
Poland	181	258	300	320
South Africa	69	119	233	300
USA	581	842	1,340	2,400
USSR	614	851	1,100	1,800
Total	2,233	3,401	5,161	8,095

[a]Included because of its importance in international trade. Source: WEC, Ref. 1.

Table 28: *OECD energy demand and supply and world oil trade Reference case: alternative nuclear sub-cases (MTOE)*[2]

	1976	*1985*	*1990*	*2000*
OECD demand				
Solid fuel (low nuclear – high nuclear)	708	993	1199–1160	1472–1313
Oil and NGL (low nuclear[a] – high nuclear)	1909	2486–2458	2756–2702	3154–3079
Natural gas	681	798–797	852–850	865–857
Nuclear (low–high)	88	295–324	447–542	925–1172
Hydro/geothermal (low nuclear – high nuclear)	226	287	324	361–356
Others		5	14	136
Total demand	3613	4864	5592	6913
OECD indigenous supply				
Solid fuel (low nuclear – high nuclear)	687	952	1130–1091	1365–1233
Oil and NGL	597	824	777	699
Natural gas	663	659	645	549
Nuclear (low–high)	88	295–324	447–542	925–1172
Hydro/geothermal (low nuclear – high nuclear)	226	287	324	361–356
Others		5	14	136
Total indigenous supply (low nuclear – high nuclear)	2262	3022–3051	3337–3393	4035–4145
OECD net imports				
Solid fuel (low nuclear – high nuclear)	31	41	69	107–80
Oil and NGL[a,b]	1322	1566	1979–1925	2455–2380
Natural gas	24	139–138	207–205	316–308
Total net imports (low nuclear – high nuclear)	1377	1746–1745	2255–2199	2878–2768
Net oil imports (exports) by world region				
OECD[b]		1566	1979–1925	2455–2380
Centrally planned economies		–	–	–
OPEC[c]		(1715)	(1685)	(1545)
Non-OPEC developing countries and others		106	136	350
World potential excess demand (supply)		(43)	430–376	1260–1185

[a]Includes bunkers.

[b]For 1985 assumes IEA Group Target of 26 Mb/d (excluding bunkers) achieved through intensified conservation and supply expansion programmes. Without the constraint imposed by the target the balance would yield net oil import demands of 33.2 Mb/d for the lower and higher nuclear sub-cases respectively.

[c]OPEC production is assumed to be 38.5 Mb/d in 1985, 37.9 Mb/d in 1990 and 40 Mb/d in 2000. These assumptions were adopted after an analysis of each country's reserve base, existing and planned capacity, economic development, and oil revenue needs. Much doubt about the economic bounds on optional production was not resolved, and at least a ± 10 per cent band of uncertainty exists around these estimates. Domestic demand plus bunkers sold by OPEC countries are projected to be 4.2 Mb/d in 1985, 5.1 Mb/d in 1990 and 9.1 Mb/d in 2000.

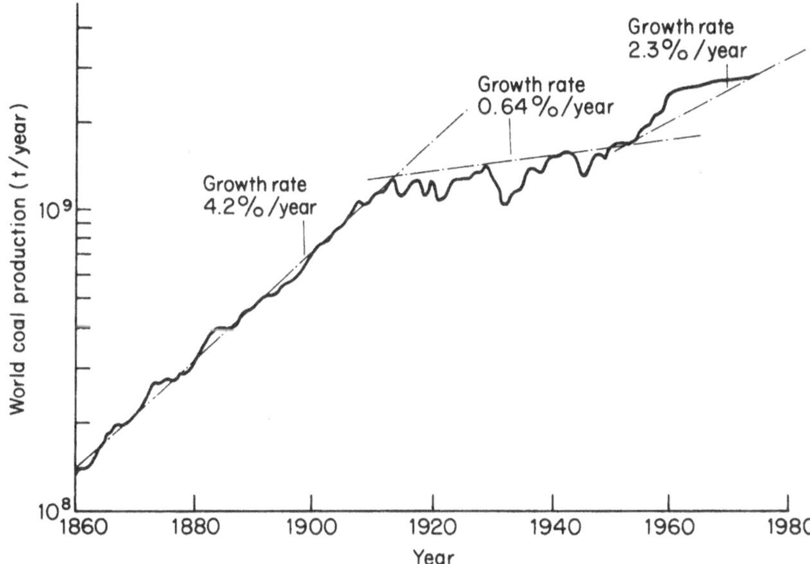

Figure 23: *World production of coal (semilogarithmic scale). (Source: WEC)*

The overall coal (solid fuel) demand is also put as increasing from 708 MTOE to between 1313–1472 MTOE. This figure can be broken down according to the different uses to which the coal is put – electricity generation, other uses and metallurgical use (Table 29).

Table 29: *Present and future patterns of use within the OECD*

	1976		*2000*	
	MTCE	*%*	*MTCE*	*%*
Electricity generation	607	60.5	1438	68.0
Other coal use	120	12.0	290	14.0
Metallurgical	275	27.5	378	18.0
	1002	100.0	2106	100.0

The basis for the year 2000 forecasts is given below.

(i) *For Electricity Production*
The total electricity supply and demand is assumed to grow by 4.6% per year from 1976 (4757 TWh) to 1985 (7153 TWh), by 3.9% per year from 1985 to 1990 (8669 TWh) and 2.9% per year from 1990 to 2000 (11,518 TWh). As in the

case of total energy demand the slow down is due to projected reductions in economic growth rates and increases in energy savings.

The fuel use for electricity generation in OECD for the low nuclear case is then shown in Figure 24.

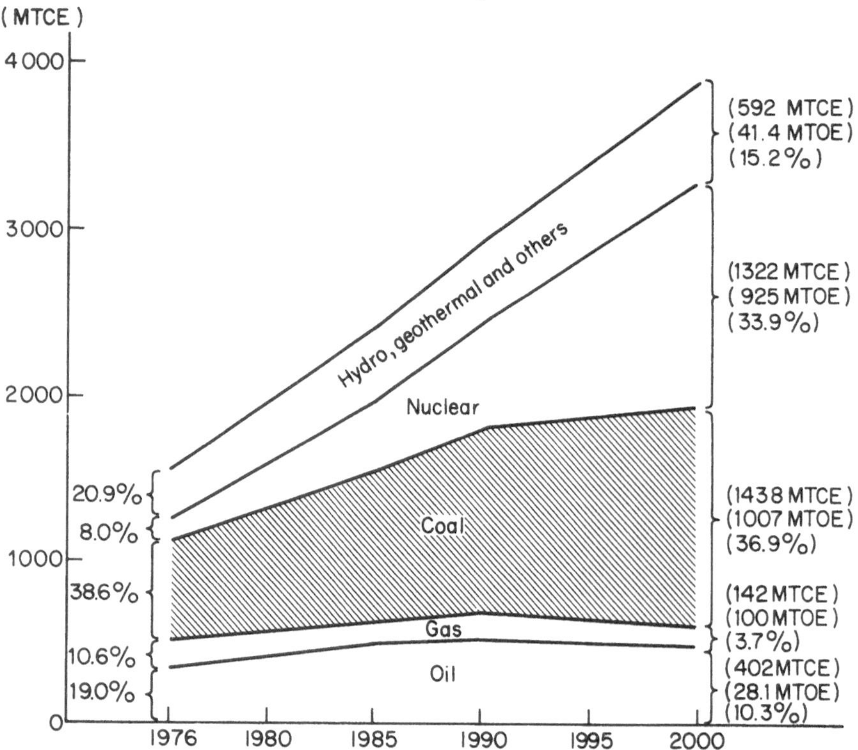

Figure 24: *Fuel used for electricity generation in the OECD*

As the share of nuclear power is projected to grow from 8.0% in 1976 (88 MTOE) to 33.9% in the year 2000 (925 MTOE) the shares of each of the other forms of energy, including coal, diminishes. But in absolute terms coal usage is to expand from 607 million tons (425 MTOE) in 1976 to 1,438 million tons (1,007 MTOE) in 2000.

(ii) *For Other Coal Use*

The demand for coal outside electricity generation is in domestic/commercial sector, industrial use, transformation into synthetic gas or liquids and non-energy use as a feedstock or raw material.

The domestic/commercial use of coal is now small, taking only 3% of the total energy consumption of these sectors in 1976, and it will probably continue to shrink largely owing to its inconvenience for small customers. On the other hand this convenience factor could be outweighed if there are severe shortages coupled with higher prices for the alternatives of oil or gas. There could also be a possible expansion of district heating networks from central coal fired stations.

In 1976 the industrial market accounted for 25% of total energy requirements in the OECD countries and of this the share of coal (thermal and metallurgical) was 20%. The OECD report suggests that there is considerable potential for increased use of coal in the industrial sector where the cement industry is given as an example. In addition once fluidized bed combustion becomes a commercial reality the attractiveness of coal for small scale industrial uses will be considerably enhanced.

(iii) *For Metallurgical Use*

The estimate of future demand for metallurgical coal is subject to uncertainty. Steel consumption per unit of GDP is falling in the advanced industrial countries and at the same time the proportion of steel produced by direct reduction from scrap without the use of coke has been increasing (although this would increase electricity consumption). The amount of coke to produce one ton of pig iron has been declining with improved production techniques. It is then recognised that the figure of 378 million tons by the year 2000 may be high.

THE ENLARGED COAL CASE

It is now widely accepted that by the year 2000 the potential world demand for oil could exceed the available supply by something between 1,000–2,000 million tons/year. This "gap" can only be closed by reducing demand or expanding supply. The "enlarged coal case" considers the potential of coal in meeting energy demand above that taken in the "reference case" by following an active policy of substituting coal for oil. This policy was endorsed by the Governing Board of the IEA meeting at Ministerial level in Paris, 21–22 May, 1979, when the "Principles for IEA Action on Coal" were adopted. The Ministers recognised the serious nature of the energy crisis and the risk that within the decade of the 1980s this could have severe economic, social and political repercussions throughout the world and in the public statement issued after the meeting concluded that:

> "Conventional oil supplies will not be available at reasonable prices and in sufficient quantities to meet the growing needs of the world during the medium and longer term.
>
> Coal, nuclear power and energy conservation are principal energy options with major potential for reducing oil dependence and thus improving the overall energy position in the medium term."

The Ministers adopted a number of "Principles for Coal Policy". Under these the IEA countries will, as common objectives expand:

(a) the use of coal as an alternative fuel;
(b) the production of coal to meet this increased demand;
(c) international trade in coal to meet increased demand.

To provide long term reliability for investments, to reduce the uncertainties and to improve the conditions under which coal use, production, and trade can develop, a number of actions and policies are proposed. These include measures to ensure that there is a reasonable balance between energy requirements and environmental standards for existing facilities so that energy costs are not disproportionately increased in relation to environmental benefits. Advanced methods for coal-mining, transport, combustion (particularly fluidised bed combustion, with improved emissions control and the disposal and utilisation of solid wastes), and conversion into other fuels will be given high priority in research and development programmes.

The use of oil for electricity generation is to be minimised. New or replacement base-load oil fired stations will be discouraged and oil will be progressively confined to middle and peak load generation. A maximum use will be made of fuels other than oil in dual-fired stations. On the other hand, the use of large coal-fired boilers when planning new industrial parks, district heating and co-generation projects will be encouraged.

To assist international trade, transport systems for both inland transportation and sea-going carriers, port facilities, and other infrastructure where necessary to handle much larger volumes of coal will be developed. The IEA countries both as producers and consumers will seek to promote long term contracts so as to achieve stability of supply and once these are in force they will not be subject to government interference except in case of extreme need.

On the supply side, countries with the potential for large increases in coal production, in particular Australia, Canada and the United States, will expand their coal production facilities and infrastructure to permit increased domestic use of coal as well as exports consistent with economic and social costs. Other significant coal producing countries, in particular the United Kingdom and Germany, while securing the level of their coal production required by energy, social and regional policies, will accept imported coal rather than oil to meet demand in excess of this level.

These policies are already being pursued. On the supply side the major oil companies are diversifying into the coal industry. In 1978 Shell reported an international coal trade of 1.6 million tonnes, about twice the 1977 figure which had led to the purchase of two coal ships. Coal mining development was being undertaken by Shell in Australia, USA, Canada and S. Africa (where the Rietspruit mine will eventually produce 6 million tonnes for export). In the US Shell Oil coal sales were 3 million tonnes. Exploration and feasibility studies were being carried out in Australia, Botswana and Swaziland. BP similarly is increasingly developing its coal interest which is now the major area of investment outside the traditional oil, tanker, gas and chemical operations. BP's production is primarily in Australia, with 6 million tonnes of coal exports in 1978 and in S. Africa. On the user side, in Japan the semi-government Electric Power Development Company which is promoting an expansion of coal fired electricity generation is negotiating to extend its stake in Australian coal mining companies by making the acquisition of a share in the capital a precondition of a long term purchasing contract.

The IEA is also promoting a number of international studies to facilitate the use of coal and to reduce the environmental pollution of coal burning. These include projects to develop the combustion of mixtures of coal and oil for use in power stations and industrial process heat boilers which would otherwise burn only oil. Fluidised bed combustion is also being studied. This is not only a more efficient method for coal burning, but since the combustion temperatures are relatively low (about 800–900°C) the amount of noxious nitrous oxides produced and discharged in the flue gas is reduced. By blending limestone or dolomite with the coal the emission of sulphur dioxide can also be reduced. This makes it possible to burn both high sulphur and low grade coals.

The proposals for the enlarged coal case, however, despite the activity noted above seem extremely modest compared with the magnitude of the task. For the low nuclear case they amount to an increase of only 500 MTCE (353 MTOE) over the reference case of 2,106 MTCE by the year 2000. But at over 2,500 MTCE it does represent an increase of 2½ times the 1976 level.

There are several obstacles in the way of a greater expansion in the short term. In the electricity sector it is not practicable to consider the complete replacement of existing oil fired stations; the most that can be expected is early retirement, or dropping down in load factor order of merit. The enlarged coal case would correspond to replacing about one-third of the oil used for electricity generation by coal by the year 2000.

For "other uses" the enlarged case figure for the year 2000 is increased from 290 MTCE to 614 MTCE. This seems an ambitious target bearing in mind the consumer resistance to the inconvenience of coal. On the other hand much more could be done to expand synthetic fuel production from coal. It has been pointed out that the technology for this was developed in the last war, in both Germany and in the UK at ICI Billingham. In Germany, the Lurgi company has claimed[2] that a plant for converting German brown coal to methanol would be competitive with oil at a price of $23/barrel. By the end of 1979 the oil price had already reached $25/barrel with up to $50/barrel being paid on the Rotterdam spot market. The response of the South African Government to the cutting off of supplies from Iran by immediately putting in hand a large expansion of their existing oil from coal SASOL plant should also be taken as an example. The Fluor Corporation of California which is managing the construction of the new SASOL plants has said[3] that a similar plant with an output of 58,000 barrels/day of oil products from 40,000 tonnes coal/day could be built close to strip mines in Wyoming for a capital cost of $3.6 billion. The cost of oil from such a plant is estimated at the equivalent of $46 per barrel of crude oil and could be in operation in five years time. While this cost is appreciably higher than the Lurgi claim it could be regarded as setting a ceiling to increases in oil price. The potential of oil-from-coal in providing the irreplaceable use of oil for the internal combustion engine should not be underestimated if serious oil shortages develop. It could then be regarded as the premium use of coal with first claim on coal production.* As nuclear power takes over an increasing share of electricity generation more coal

* Some steps are now being taken in this respect. The German Government has announced plans to invest some $6.5 billion in building 14 coal processing plants during the next decade. The European Commission has forecast a consumption of 25 million tons of coal by 2000 for coal liquefying and gasifying plants.

could be released for synthetic oil production.

Another limitation, which the IEA proposals are intended to reduce, is in developing an international export trade in coal. At present most of the major coal producing countries consume the greater part of their production within their own countries. On average the export/production ratio is below 10% (Figure 25).

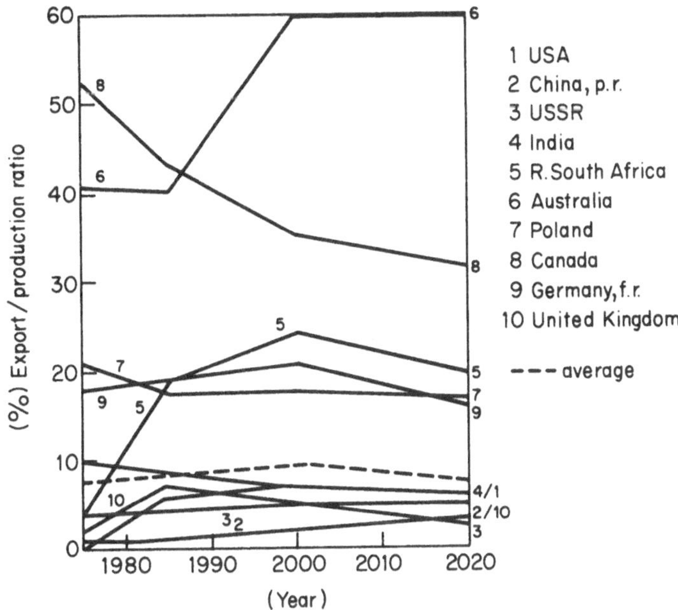

Figure 25: *Development of the export/production ratio (%) of main coal producing countries. (Source: Ref. 5)*

Only Canada, Australia, Poland and S. Africa seem likely to develop as significant exporters of coal. This could change if the energy shortages pushed up the price of coal, but this would require time to build export terminals, organise bulk carriers for shipment and to increase production above that required for domestic use so as to have a surplus available for export. The possibility of a country, such as China, a major world producer which at present only exports 1% of its production, developing as a coal supply country should not be ruled out (Figure 26 and Table 30).

It is probable that the expansion of coal production of the Eastern Bloc countries would take place in a similar way to those of OECD. On the basis (as at present) that they produce one half of the world total, this would give a world production in

Table 30: *Data of production and export of the main coal producing countries (MTCE)*

Country	1975			1985			2000			2020		
	Produc-tion	*Export*	*Export (%)*	*Produc-tion*	*Export*	*Export (%)*	*Produc-tion*	*Export*	*Export (%)*	*Produc-tion*	*Export*	*Export (%)*
Australia	69	29	42	150	60	40	300	180	60	400	240	60
Canada	23	12	52	35	15	43	115	40	35	200	65	32
China (People's Republic)	349	3	1	725	7	1	1200	30	2	1800	50	3
Fed. Republic of Germany	126	23	18	129	25	19	145	30	21	155	30	19
India	73	–	–	135	7	6	235	13	7	500	32	6
Japan	19	–	–	20	–	–	20	–	–	20	–	–
Poland	181	39	21	258	45	17	300	50	17	320	50	16
South Africa (Republic)	69	3	4	119	23	19	233	55	24	300	60	20
United Kingdom	129	2	2	137	10	7	173	10	6	200	10	5
USA	581	60	10	842	68	8	1340	90	7	2400	145	6
USSR	614	26	4	851	37	4	1100	50	5	1800	60	3
Other countries	360	2	1	483	6	1	619	34	6	751	46	6
Total	*2593*	*199*	*7.7*	*3884*	*303*	*7.8*	*5780*	*582*	*10.1*	*8846*	*788*	*8.9*

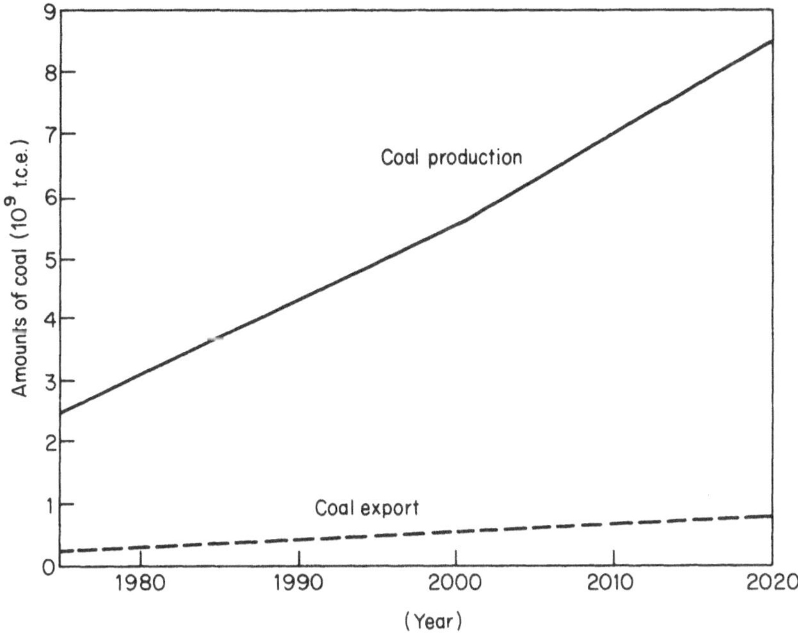

Figure 26: *Development of coal production and coal export (world). (Source: Ref. 5)*

the year 2000 of just over 5,000 MTCE – a figure which conforms with the WEC estimate. But even if this figure is achieved it may do no more than cover about one-half of the expected "energy gap" of the year 2000. Too many countries looking at their own requirements in isolation now seem to believe that they can always fall back upon supplies of imported coal to meet their energy deficit. This is not so. The scramble for coal supplies has already started and it is unlikely that there will be enough to go round. The world must either reduce demand or find an additional source of energy.

REFERENCES

1. *World Energy: Looking Ahead to 2000*, Conservation Commission of World Energy Conference, 1978.
2. *Steam Coal: Prospects to 2000*, International Energy Agency.
3. *Financial Times*, 15 June 1979.
4. *Financial Times*, 19 September 1979.
5. *World Energy Resources 1985–2020*, World Energy Conference 1978, IPC Science and Technology Press, Guildford.

5
Nuclear power

Chapters 3 and 4 show that with probable restrictions on the output of oil and despite an accelerated expansion of the coal industry, coupled with draconian measures of energy conservation, most studies indicate that substantial nuclear power output is required by the year 2000 if sufficient energy is to be available to maintain a modest growth in the industrial countries, to fuel the more rapid growth of the developing countries and to ease the burden of poverty in the under-developed countries.

In considering how far and how fast nuclear power capacity might grow to meet this demand it is useful to review the rapid development of the industry over the past 37 years, since the first man-made self-sustaining nuclear chain reaction was obtained by Professor Fermi and his co-workers with the now famous "pile" in the squash court of Chicago University on 2 December 1942.

The development of nuclear energy was pushed initially by the military programmes. The plutonium production piles at Hanford USA, Windscale UK and Marcoule in France were all built to produce plutonium for nuclear weapons while the first power applications came with the nuclear submarine propulsion units for the USA. The application for civil electricity generation followed quickly. In the UK once-through air cooled graphite piles of Windscale led to the closed cycle, carbon dioxide cooled reactors of Calder Hall; the first of which came into operation in 1956. In the USA the first civil nuclear power station, Shippingport, started up in 1957. These stations were still more of the nature of prototypes rather than fully commercial stations and the nuclear power era is generally accepted as being ushered in with the order for the Oyster Creek nuclear power station in 1963 given to General Electric in the USA and the Magnox reactors of the UK. Oyster Creek came into operation in 1969, the Magnox reactors in 1962. These stations were bought by electricity companies because, for the first time, commercial tenders for nuclear power plant showed that the electricity they would generate was cheaper than from contemporary coal or oil fired stations.

The following years showed a rapid expansion of nuclear power

Table 31: *Nuclear power. State of programmes in various countries (MW(e) net). Situation at 1 January 1979*

Geographical area	Completed		In operation[a]		Under construction or ordered		Total	
	MW(e)	No. reactors	MW(e)	No. reactors	MW(e)	No. reactors	MW(e)	No. reactors
USA	288	7	52,145	73.00	137,881	124.00	190,314	204.00
W. Europe								
Austria[b]	0	0	0	0.00	692	1.00	692	1.00
Belgium	0	0	1,339	4.00	3,825	4.00	5,164	8.00
Finland	0	0	1,080	2.00	2,080	3.00	3,160	5.00
France	73	2	6,933	14.25	30,435	31,20	37,441	47.45
Italy	0	0	1,397	4.00	4,040	5.00	5,437	9.00
Yugoslavia	0	0	0	0.00	615	1.00	615	1.00
Luxembourg	0	0	0	0.00	1,247	1.00	1,247	1.00
Holland	0	0	499	2.00	0	0.00	499	2.00
UK	14	1	7,893	33.00	3,674	6.00	11,581	40.00
W. Germany	125	2	8,152	15.00	17,422	16.15	25,699	33.15
Spain	0	0	953	2.75	13,168	14.00	14,121	16.75
Sweden	140	2	3,722	6.00	5,720	6.00	9,582	14.00
Switzerland	7	1	1,926	4.00	2,683	2.65	4,616	7.65
Total W. Europe	359	8	33,894	87.00	85,601	91.00	119,854	186.00
E. Europe								
Bulgaria	0	0	880	2.00	880	2.00	1,760	4.00
Czechoslovakia	0	0	590	2.00	1,320	3.00	1,910	5.00
Poland	0	0	0	0.00	440	1.00	440	1.00
E. Germany	0	0	1,390	4.00	440	1.00	1,830	5.00
Rumania	0	0	0	0.00	1,040	2.00	1,040	2.00
Hungary	0	0	0	0.00	880	2.00	880	2.00
USSR	0	0	10,983	30.00	25,765	28.00	36,658	58.00
Total E. Europe			13,753	38.00	30,765	39.00	44,518	77.00
Other countries								
Argentina	0	0	319	1.00	600	1.00	919	2.00
Brazil	0	0	0	0.00	3,214	3.00	3,214	3.00
Canada	0	0	5,516	11.00	9,527	14.00	15,043	25.00
Taiwan	0	0	1,208	2.00	3,716	4.00	4,924	6.00
S. Korea	0	0	564	1.00	3,034	4.00	3,598	5.00
Philippines	0	0	0	0.00	626	1.00	626	1.00
Japan	0	0	13,480	22.00	10,081	12.00	23,561	34.00
India	0	0	580	3.00	1,080	5.00	1,660	8.00
Mexico	0	0	0	0.00	1,290	2.00	1,290	2.00
Pakistan	0	0	125	1.00	0	0.00	125	1.00
Iran[c]	0	0	0	0.00	8,992	8.00	8,992	8.00
S. Africa	0	0	0	0.00	1,844	2.00	1,844	2.00
Total other countries	0	0	21,792	41.00	44,004	56.00	65,796	97.00
Total world	647	15	121,584	238.00	299,251	311.00	420,482	564.00

[a] Those reactors about to produce in 1978 have also been taken into account.
[b] The completed Austrian station at Zwentendorf has not been allowed to operate, as a consequence of the national referendum of 5 November 1978.
[c] The Iranian programme has now been abandoned.
Source: Ref. 1.

as more manufacturers and fuel supply companies entered this growing market. By 1 January 1979 there were 223 power reactors in operation in 22 countries with a total capacity of just over 120 GW. When the plants now under construction or on order are included the figure for nuclear power capacity increases nearly fourfold to 420 GW from 549 stations in 31 countries. Table 31 lists the nuclear power reactors which are in operation, those ordered or under construction, and the small number, mainly experimental or early prototype reactors which have now been shut down.

Table 32 ranks the countries according to the nuclear-capacity in operation.

Table 32: *Nuclear power in operation.*[a] *Situation at 1 January 1979*

	Country	*MW(e)net*
1.	USA	52,145
2.	Japan	13,480
3.	USSR	10,893
4.	W. Germany	8,152
5.	UK	7,893
6.	France	6,933
7.	Canada	5,516
8.	Sweden	3,722
9.	Switzerland	1,926
10.	Italy	1,397
11.	E. Germany	1,390
12.	Belgium	1,339
13.	Taiwan	1,208
14.	Finland	1,080
15.	Spain	953
16.	Bulgaria	880
17.	Czechoslovakia	590
18.	India	580
19.	S. Korea	564
20.	Holland	499
21.	Argentina	319
22.	Pakistan	125
	Total world	121,584

[a] Those reactors about to produce in 1978 have also been taken into account.
Source: Ref. 1.

Table 33 shows that when stations under construction or on order are included the number of countries with nuclear programmes goes up from 22 to 33 and the nuclear power capacity increases from 122 GW to 420 GW. In this ranking the United

States retains its position as the leading nuclear country, but is now followed by France, which jumps from 6th to 2nd place, and the USSR.

Table 33: *Nuclear power in operation, under construction or ordered.*[a] *Situation at 1 January 1979*

	Country	MW(e) net
1.	USA	190,026
2.	France	37,368
3.	USSR	36,658
4.	W. Germany	25,574
5.	Japan	23,561
6.	Canada	15,043
7.	Spain	14,121
8.	UK	11,567
9.	Sweden	9,442
10.	Iran[b]	8,992
11.	Italy	5,437
12.	Belgium	5,164
13.	Taiwan	4,924
14.	Switzerland	4,609
15.	S. Korea	3,598
16.	Brazil	3,214
17.	Finland	3,160
18.	Czechoslovakia	1,910
19.	S. Africa	1,844
20.	E. Germany	1,830
21.	Bulgaria	1,760
22.	India	1,660
23.	Mexico	1,290
24.	Luxembourg	1,247
25.	Rumania	1,040
26.	Argentina	919
27.	Hungary	880
28.	Austria[b]	692
29.	Philippines	626
30.	Yugoslavia	615
31.	Holland	499
32.	Poland	440
33.	Pakistan	125
	Total world	419,835

[a]Those nuclear power stations definitely out of use (699 MW(e) of which 304 (MW(e) is in the US and 395 MW(e) in W. Europe) have not been taken into consideration.
[b]The Iranian programme has been abandoned; the Austrian and Luxembourg plans are in abeyance.
Source: Ref. 1.

Nuclear power is already making a substantial contribution towards electricity supply. As Table 34 shows it is now more than 20% of total electricity in Sweden and Belgium, and in these countries, with Canada, USA and Switzerland the average consumption per capita of nuclear electricity is over 1000 kWh/year.

Table 34: *Contribution of nuclear energy to the total production of electric power*

Country	Population (in millions)	Total power production 1978 (billion kWh)	Total power production 1978 kWh per capita	Nuclear power production 1978 (billion kWh)	Nuclear power production 1978 kWh per capita	Contribution of nuclear power to total (%)
USA	215.9	2,446.5	11,332	293.6	1,360	12.0
W. Europe						
Belgium	9.7	48.5	5,000	10.2	1,052	21.0
Finland	4.7	35.3	7,511	3.2	681	9.1
France	52.9	217.0	4,102	33.7	637	15.5
Italy	56.6	167.4	2,958	4.4	78	2.6
Holland	13.7	59.4	4,336	4.0	292	6.7
UK	56.0	269.0	4,804	39.1	698	14.5
W. Germany	61.8	331.0	5,356	32.6	528	9.8
Spain	36.1	113.0	3,130	6.8	188	6.0
Sweden	8.2	120.3	14,671	26.1	3,183	21.7
Switzerland	6.4	49.7	7,766	8.3	1,297	16.7
E. Europe						
USSR	257.9	1,202.0	4,661	44.0	171	3.7
Other countries						
Argentina	26.0	44.6	1,715	2.9	112	6.5
Canada	23.0	331.0	14,391	33.1	1,439	10.0
Japan	110.0	847.9	7,708	50.9	463	6.0
S. Korea	35.9	23.0	641	2.3	64	10.0
India	610.0	115.0	189	2.3	4	2.0

Source: Ref. 1.

This rapid growth of nuclear power capacity has been brought about through the development of a substantial nuclear industry on a world wide scale. Although this industry has been dominated by the two principal American manufacturers, Westinghouse and General Electric, who pioneered the two principal reactor systems, the Pressurised Water and Boiling Water reactors respectively, other reactor suppliers, in Europe, Japan and the US, are now taking an increasing share of the market.

The dominance of the light water reactors is clearly shown in Table 35. It is only the gas cooled graphite reactors built by the UK and France, the Canadian heavy water reactors and the Russian water cooled graphite reactors which have offered an alternative reactor system. But Russia also builds PWR stations and in France the gas cooled line has been abandoned in favour of the PWR.

Table 35: *Contributions of the different reactor types in nuclear power ordered. Situation at 1 January 1979*

Geographical area	Type of reactor													
	PWR[a]		BWR[b]		PHWR[c] e SGHWR[d]		GCR[e] e AGR[f]		FBR[g]		Others		Total	
	MW(e)	%	MW(e)	%	MW(e)	%	MW(e)	%	MW(e)	%	MW(e)	%	MW(e)	%
USA	132,113	69.4	56,305	29.6	–	–	–	–	430	0.2	1,466	0.8	190,314	100
W. Europe	75,931	63.4	26,875	22.4	154	0.1	14,221	11.9	2,013	1.7	660	0.5	119,854	100
E. Europe	25,160	56.5	50	0.1	600	1.3	–	–	762	1.7	17,946	40.4	44,518	100
Other countries	27,756	43.7	18,429	28.0	17,746	27.0	159	0.2	300	0.5	415	0.6	65,796	100
Total world	261,960	62.3	101,650	24.2	18,500	4.4	14,380	3.4	3,505	0.8	20,487	4.9	420,482	100

[a]Pressurised water reactor
[b]Boiling water reactor
[c]Pressurised heavy water reactor
[d]Steam generating heavy water reactor
[e]Gas cooled reactor
[f]Advanced gas cooled reactor
[g]Fast breeder reactor
Source: Ref. 1.

The principal reactor manufacturers and the type of reactor they offer are shown in Table 36 which ranks them according to the number of orders they have received.

Table 36: *Orders received by the principal manufacturers. Situation at 1 January 1979*

	Manufacturer	*Nationality*	*MW(e)*
1.	Westinghouse	USA	87,059
2.	General Electric	USA	70,193
3.	USSR	USSR	45,758
4.	Kraftwerk Union	W. Germany	36,976
5.	Framatome*	France	36,324
6.	Combustion Eng.	USA	34,206
7.	Babock & Wilcox	USA	27,796
8.	AECL	Canada	18,277
9.	NPC (and predecessors)	UK	11,581
10.	Toshiba†	Japan	8,709
11.	ASEA ATOM	Sweden	8,280
12.	MAPI*	Japan	7,869
13.	Westinghouse (Europe)	Belgium	6,518
14.	Ansaldo Mecc. Nucleare†	Italy	2,880
15.	Hitachi†	Japan	2,279
16.	Elettronucleare italiana*	Italy	2,000
18.	Other manufacturers		13,787
	Total world		420,482

* Licence Westinghouse.
† Licence General Electric
Source: Ref. 1.

This is however not a static industry and continuous development and evolution takes place. In the UK for instance the five original gas graphite reactor groups have gradually been reduced by mergers and withdrawals to one. But at the end of 1978 a new Anglo-American group of Rolls Royce, Northern Engineering Industries and Combustion Engineering emerged to offer PWR.

Although the original Light Water reactor designers have a number of licencees these have in some cases been able to evolve and develop their own reactor design; KWU was formed by a merger between Siemens, originally holding a Westinghouse PWR licence, and AEG which holds a GE BWR licence.

For the boiling water reactor there is a growing international co-operation on improved design between GE, its Japanese and Italian licencees, and the Swedish ASEA ATOM which has developed an independent BWR design.

Spain which does not have a national reactor manufacturing company has however made substantial investment to establish a national capability to manufacture the primary reactor components and to supply nuclear fuel services.

There are also increasing transfers of nuclear technology from the industrial to the developing countries. One example is Brazil which is working closely with the German industry on a programme which is planned to build up to 10,000 MW of nuclear power by 1990. This is being done through a number of joint ventures through Brazilian–German companies which cover the entire nuclear field including the manufacture of reactor components as well as the fuel cycle industry so that Brazil will eventually attain an independent nuclear manufacturing capability within the shortest practical time. Korea, another developing country with a very high growth rate, averaging 9.5%/year for GNP over the period 1961 to 1976 with a growth in electricity demand over the same period at an average of 18.3%/year, is also seeking to develop an indigenous nuclear industry by licence agreements and joint ventures which are applied to nuclear power plans now in progress.

Although there is a tendency to consider nuclear industry as being dominated by the reactor design and supply companies who give their name to the reactor systems this is not a complete picture in that it does not adequately reflect the contribution of major component suppliers; of the many sub-contractors; and of the architect/engineering companies who often are given the responsibility for the overall design and construction of the complete power station of which the reactor is but a part. This can be seen from Table 37 which gives an approximate breakdown of costs for Pressurised Water Reactor nuclear power stations (excluding interest during construction, insurance, taxes, switchyard transmission lines, cost of land and the nuclear fuel). The reactor itself or as it is sometimes called, the Nuclear Steam Supply System accounts for only 15.7% of the total cost of the power station.

The nuclear steam supply system – basically the reactor and its main components, pressure vessel, steam generators and pressuriser, steam separators and driers, primary circuit pumps and piping – is often, but not exclusively, supplied by the reactor vendor. There are, for instance, in addition to the principal reactor suppliers a number of independent reactor pressure vessel manufacturers in US, Europe and Japan who can supply both BWR and PWR vessels, including the steam generator shells and pressurisers.

The reactor vendors are, however, responsible for the overall design of the NSSS and give the performance guarantees.

Table 37: *Costs of a PWR nuclear power station*

Cost item	*US $ million (1973 levels)*	*%*
Nuclear steam supply system and auxiliaries (NSSS)	48.5	15.7
Main turbine-generator set and auxiliaries (T/G)	40.9	13.2
Balance-of-plant equipment (BOP) – mechanical, instrumentation, electrical	45.0	14.6
Field erection labour (FEL) for NSSS, T/G, BOP	31.9	10.2
Civl-structural work (C/S) including labour and materials	42.3	13.7
Field indirect costs (FIC) – construction facilities and equipment, site services, plant start-up	31.4	10.1
Engineering – procurement – construction management services (EPCM)	42.6	13.7
Contingency allowance (CA) approximately 10% of the above items	27.4	8.8
Total cost of the above items	310.0	100.0

Source: Bechtel

The reactor vendor also supplies the first fuel charge, and often offers subsequent fuel reloads at the time of the reactor sale. There are, however, independent fuel supply companies which will supply replacement fuel for all reactor systems. These include Exxon and BNFL. Since cost of a full fuel load is about 10% of the whole nuclear station it corresponds to almost two-thirds of the cost of the Nuclear Steam Supply System. This emphasises the importance of continuing fuel supply business to the reactor vendors.

Following the lead given by the United States reactor licensing procedures the plant equipment and components are divided into three classes to meet the requirements of the very strict standards of quality assurance procedures. The Class I components – those components whose failure may cause or contribute to a nuclear accident – have to pass rigorous approval tests and the companies that manufacture them are subject to control and inspection. In the United States those companies whose manufacturing methods and procedures meet these standards are certified by being authorised to use the ASME "N" stamp. Such items are most likely to be found in the reactor itself, as part of the nuclear steam supply system. The reactor vendors then usually manufacture themselves some of the more critical items – the reactor internals, control rods and drive systems and often the reactor pressure vessel, steam generators and pressuriser (for the PWR). There are however a number of independent specialist suppliers of pressure vessels, and other primary circuit components including valves, pumps and pipework.

Of the main reactor vendors only the US companies, Westinghouse, Babock and Wilcox and Combustion Engineering and Mitsubishi, the Japanese licencee of Westinghouse, have an "in-house" pressure vessel capability.*

The turbine-generator is another key section of the power plant. This is sometimes the subject of a separate contract, particularly in the case where the reactor vendor is not a turbine manufacturer as for Framatome, Babcock and Wilcox and Combustion Engineering, who came into the nuclear industry from the steam raising or heavy engineering end. Most of the other reactor companies developed their nuclear capability from the electrical engineering end, such as GE, Westinghouse, Kraftwerk Union, ASEA and the Japanese companies.

The balance-of-plant equipment comprises the mechanical, electrical and other systems which lie outside the main areas of the reactor and the turbine generator. Some of these are listed in Table 38.

Table 38: *Balance-of-plant equipment*

Cost elements	*Estimated (US $ million)*	%
Rotating mechanical equipment	5.1	11.3
Tanks, vessels, heat exchangers	10.4	23.2
Mechanical handling equipment	1.3	2.9
Process piping systems	12.3	27.3
Instrumentation and control	3.5	7.8
Power transformers and isophase bus	4.0	8.9
Cable systems and penetrations	1.8	4.0
Other electrical equipment	1.9	4.2
Heating, ventilating, air conditioning equipment	4.7	10.4
Total BOP equipment	45.0	100.0

Source: Bechtel

From the Quality Assurance point of view most of these items fall into Class 2, components whose failure may cause or prolong loss of output, or Class 3, components of good commercial quality. This then brings in a large number of potential suppliers of fabricated equipment and raw materials. Valves for instance can be regarded as a special product class. A typical 1,000 MW LWR will require between 5,000–7,000 valves in its pipework systems, not counting control valves and those that are supplied as part of other manufactured components. These valves range in size from ¼ to 96 inches diameter. They include all types of valve

* In France, Framatone is closely inter-linked with its pressure vessel manufacturer Creusot-Loire.

design, gate, globe, check, diaphragm, needle, plug, stopcheck or butterfly and are mostly manufactured to Class 1 or Class 2 specifications. Piping is another substantial item. It includes pressure pipework and fittings in carbon, alloy and stainless steel, tubing for heat exchangers and instruments, in brass, cupro-nickel, stainless steel and copper, utility piping in galvanised steel, ductile iron and polyvinyl chloride and also electrical conduit and fittings. As raw materials, high quality steels are required for liner plates, pre-stressing wires and tendons, fuel pool gates, airlocks and transformer cores, while monel, nickel and special alloys are used for reactor components. The fuel cans for the LWR are normally manufactured from reactor grade zirconium; this is specially refined to separate the hafnium – a neutron poison – which always occurs together with zirconium.

The construction of a nuclear power station requires the participation of a very large number of industrial companies. One buyers guide of "Nuclear Products Materials and Services" lists some 400 items and over 1,800 different suppliers. In Germany Kraftwerk Union have said that about 700 different companies are involved in the construction of a nuclear power station. Many of these are small companies – more than 70% employ less than 200 workers – but they are often specialised with a high degree of expertise.

It is then misleading to make sweeping generalisations about the nuclear industry which is built up from so many different companies with diverse interests and where the NSSS vendor who is often "credited" with the sale may only be responsible for a minor part of the complete station. Many of the companies involved have other interests in power engineering and their emergence as "nuclear companies" is only a reflection of their belief that this is the most economic and environmentally advantageous form of power generation. These companies could equally well – and in many cases do – apply their capability to the construction of fossil fired and hydro power stations.* This is particularly the case for the component suppliers where valves, pumps etc. are required for all industrial plant: the major difference is in the higher quality demanded for the nuclear use where the requirements of Quality Assurance systems, clean room manufacture for some components have often led to suppliers setting up special facilities for nuclear manufacture to Class 1 standards rather than try to produce equipment to two levels of quality in the same factory.

*A number of "nuclear" companies are also involved in solar power development.

The reasons for the rapid growth of nuclear power as shown in Table 31 are primarily of an economic nature. Nuclear power generates electricity at a lower cost than other fuels and also reduces dependence on fuel imports. The first point is of prime importance to the individual utility companies, the second is often seen as a matter of national economic policy.

The cost structure of nuclear power differs from conventional power in that the nuclear stations have a high initial capital cost, but low fuel costs whereas a conventional fossil fired station has lower capital cost but much higher fuel costs. Because of the high capital cost of the nuclear plant financial charges are a major element in the total cost. A paper at the IAEA Salzburg Congress in 1977[2] showed that the initial investment, taken as 500 million US dollars in 1976 currency for a 1,000 MW(e)* LWR nuclear plant, would be increased to $1,040 million with a construction period of seven years from the date of authorisation to proceed, assuming an interest during construction of 10%/year and an inflation rate of 8%/year. The initial investment then represents only 48% of the final cost, inflation 20% and interest during construction 32%. The rate at which money can be borrowed is then one of the key factors in nuclear power economics and may determine the decision, by an investor owned utility, to order a nuclear station. This can be seen for the US when the ordering rate of nuclear power stations is plotted against the prime bank rate and the US bond ratings (Figure 27), the one mirrors the other almost exactly after allowing a time lag of up to one year. The disturbances that followed the OPEC embargo/price rise at the end of 1973 with a rapid escalation of fuel prices, high inflation and the recession did not encourage the US utilities to take advantage of the drop in interest rates to 1976. In any event this was only short lived, the rates have now risen to record levels. Other countries, particularly those with national electricity boards, e.g. France and UK, have been less affected by movements in interest rates since the nuclear ordering programme is treated as a matter of national policy for which the Government provides the funds directly or backs commercial loans. With the higher interest and inflation rates now being experienced in many countries the additional costs of financing nuclear power will be even higher. This emphasises the need to reduce delays in the licensing and regulatory procedures, once construction has started, and for the regulatory authorities, the utilities and manufacturers to concentrate greater effort on shortening construction time.

* (e) = electrical

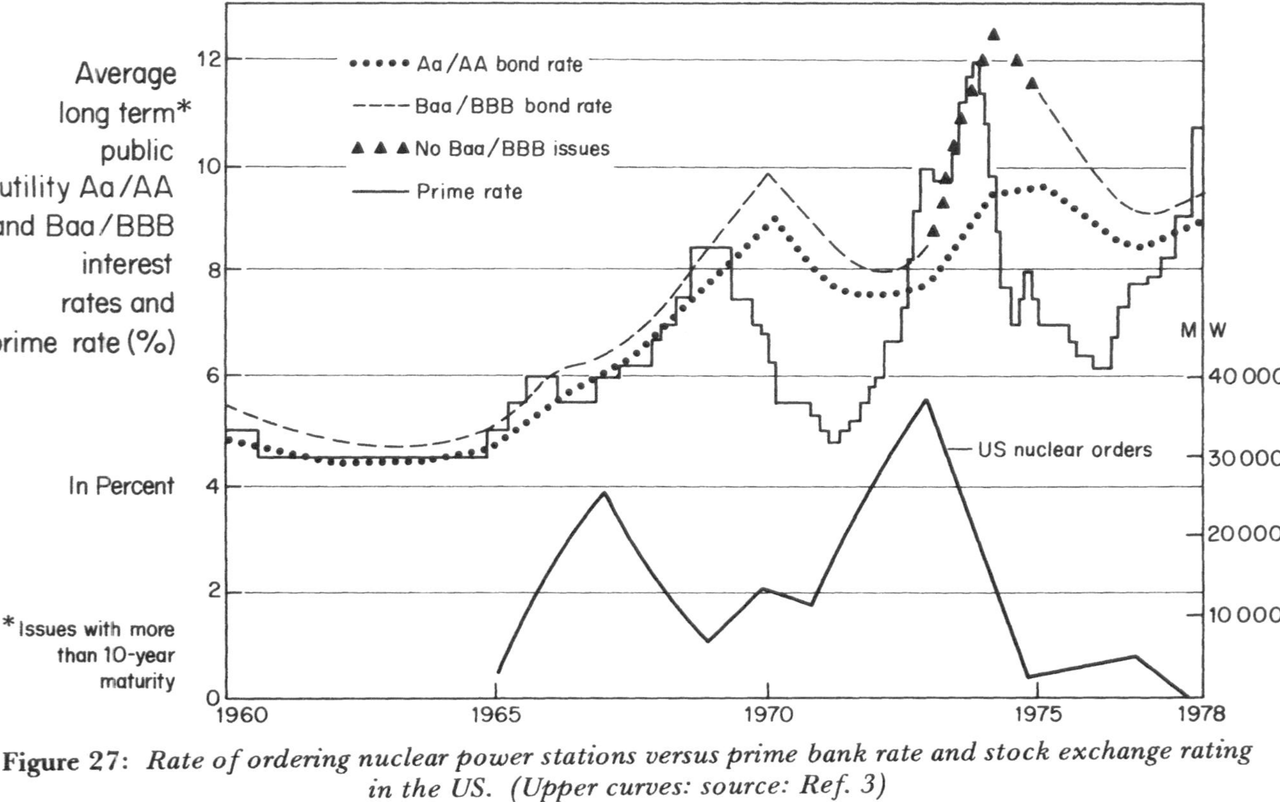

Figure 27: *Rate of ordering nuclear power stations versus prime bank rate and stock exchange rating in the US. (Upper curves: source: Ref. 3)*

Some nuclear plants have been built – in Japan and Switzerland – in four to five years. The aim should be to make this the standard rather than an exception.

The delays in construction are perhaps most remarkable in the United States where up to 12–14 years is now required for completion to commercial operations. The reasons that have been ascribed for this are[4] firstly technical factors arising from the rapid increase in plant size and a continuing lack of standardisation of plant design; a second (and probably weightier) reason is the increasing number and stringency of safety and environmental regulations, the linkage of power plant licensing to the resolution of the nuclear waste disposal issue, and the continuing fluidity and uncertainty surrounding the content and application of regulations. Of particular importance have been the effect of regulatory changes during construction and design with the "ratchetting" and "backfitting" that this entails. This 12–14 year construction time, combined with very high interest rates has temporarily halted nuclear ordering in the US.

Despite these high financing charges nuclear power maintains a differential of anything between 60–10% over electricity from fossil fired stations. The latter have also been affected by increasing capital costs and by the need for flue gas cleaning equipment to meet stricter environmental controls, but the important factor is the higher cost of fossil fuels. These will be susceptible to inflationary increases over the whole life of the plant – a period of 20–30 years – while for the nuclear plant, once it is built, inflation will only increase its competitive advantage because of its much lower fuel cost.

From the national point of view the importance of nuclear power is that it can reduce the dependence on fuel imports and the drain on the national economy in paying for such imports.

In some countries the dependence on fuel imports is very high. Table 39 gives figures for the USA and Europe.

This makes reliability of supply of utmost importance. Now that oil is being used as a political weapon to advance the policies favoured by the Arab oil producing states, world peace could be endangered by a political miscalculation as to how far a threat can be pushed. For nuclear fuel on the other hand the raw material – uranium – is a widely spread mineral and there are a number of competing suppliers offering fuel cycle services. It would, in any case, with the very much smaller quantities involved, be relatively easy for an importing country to store several years supply of nuclear fuel.

The cost to the economy of fuel imports has been less widely

Table 39: *Energy balance (MTOE), 1975*

	Total energy requirements	*Indigenous production*	*Self sufficiency %*
USA	1,690.15	1,455.96	86.0
OECD Europe	1,116.26	491.96	44.0
Austria	22.86	10.30	45.0
Belgium	41.83	6.91	16.5
Luxembourg	3.99	0.16	4.0
Denmark	17.70	0.15	0.8
Finland	22.30	6.63	30.0
France	168.06	43.29	26.0
Germany	243.49	118.92	49.0
Ireland	7.08	1.55	22.0
Italy	127.03	25.23	20.0
Netherlands	58.99	72.29	123.0
Norway	19.15	21.22	111.0
Portugal	8.20	1.65	20.0
Spain	60.68	17.29	29.0
Sweden	49.10	19.37	40.0
Switzerland	22.44	9.68	43.0
UK[a]	203.22	116.95	58.0

[a]With the development of North Sea oil, the UK is expected to reach self sufficiency for some years from 1980 onwards and even be a net exporter of energy.
Source: GHG.

Table 40: *Energy imports as a % of merchandise export earnings in 1976*

	%
Low income countries average	19
India	26
Middle income countries average	22
Korea	23
Brazil	43
Portugal	38
Greece	48
Spain	59
Industrialised countries average	24
Italy	30
UK	22
Japan	42
Austria	17
Finland	25
France	26
Australia	8
USA	30
Belgium	15
Denmark	23
Germany, Fed. Rep.	16
Canada	11
Sweden	18
Switzerland	11
Centrally planned economies	
Cuba	39
Hungary	14
USSR	4

appreciated. Yet this can be very high as is seen in the figures in Table 40, from the World Bank[5] which expresses energy imports as a percentage of merchandise export earnings.

The need for substantial nuclear power programmes in countries such as Brazil, India, Japan and even the United States is then plain. No country can for long carry the burden of paying for energy imports at such a high cost to the economy.

This problem is now hitting hardest at some of the developing countries as has been pointed out by Munir Khan, Chairman of the Pakistan Atomic Energy Authority:[6]

> "The industrialised countries developed their basic industrial infrastructure at a time when energy costs were very low – oil was priced at $1–2 per barrel and coal at $2–3 per ton as recently as the mid-1960s. But now when the LDCs have started building their industrial infrastructure and require enormous inputs of energy, the cost of energy has risen by an order of magnitude, making it extremely expensive for them to embark upon large-scale development programmes."

The countries which are now taking over the burden of producing the energy-intensive goods and materials, that is now gradually being relinquished by the West cannot afford to turn to anything but the cheapest power available, and this is often nuclear power – despite the fact that the size of station available is in some cases too large for the electricity grid systems available. This can be seen where the percentage share of nuclear electricity in some of the advanced developing countries is already equal to or even greater than for some of the industrial countries (Table 41).

Table 41: *Share of nuclear power in total electricity (%)*

	%		%
Bulgaria	18	FRG	9
India	2	Italy	2
Taiwan	10	Netherlands	6
Pakistan	2	Japan	8
Korea	7	USSR	3
Argentina	8		

To see how nuclear power can be expected to contribute to the energy requirements of the developing countries, the cases of Korea, India and Brazil are considered in more detail.

KOREA

The energy demand pattern of Korea, now rapidly emerging as an industrial country, can be taken as an example of the course that will be followed by other developing countries making the same transition. One remarkable feature is the phenomenal rate of growth of electricity demand; which has increased no less than 27-fold between 1961 and 1978, rising from 1,173 to 31,510 GWh. The installed capacity increased from 367 to 6,916 MW.

Korea does not have adequate indigenous energy resources and relies heavily on imported oil which in 1975 at 33.97 MTCE met 56.7% of energy consumption. By the year 2000 total energy demand is expected to increase to ten times the 1975 figure, with oil at 297 MTCE meeting 50.2% of the demand. Coal, of which the economically recoverable resources are put at between 500–1,000 million tons (depending on the price of oil) will be depleted over the next 25–30 years. The hydro power potential is put at 3,000 MW and it is believed that tidal power may be feasible after 1985, with a possible capacity of 4,000 MW. This then foresees nuclear power as supplying 20.7% of the primary energy demand by the year 2000, when the nuclear capacity is expected to reach 79,396 MW, approximately 66% of the electrical capacity (Table 42).[7]

A first step towards the realisation of this ambitious programme was taken when the nuclear power station Ko-Ri 1, near Pusan, a 560 MW PWR, began operation in April 1978. Two more stations are now well under way and should be in operation by 1982. These are Ko-Ri 2, a 600 MW PWR, and Wolsung 1, a 630 MW Candu heavy water reactor. These first three reactors are being built under turn-key contracts, the PWRs by Westinghouse with a British GEC turbine generator, the Candu by AECL with a Howden-Parsons turbine generator.

It is however the Government's policy to develop a complete national nuclear capability for the 1990s. For the next two stations, Ko-Ri 5 and Ko-Ri 6, both 900 MW PWRs, now starting construction for operation in 1984 and 1985, the lead was taken by the national utility, Korea Electric Co. who appointed Bechtel to carry out the design and project engineering with the major sections of the plant, the NSSS and initial fuel coming from Westinghouse and the turbine generators from GEC. The local content of supply has then increased from 8% for Ko-Ri 1, to 12% for Ko-Ri 2, while for Ko-Ri 5 and 6 it will rise to 24% of equipment and materials, mainly class 3 components in the balance of plant area. Under the agreement with Bechtel, KNE (Korea

Table 42: *Energy consumption structure (1,000 MTCE)*

		1975	*1981*	*1986*	*1991*	*1996*	*2000*
Coal		18,460	28,600	47,023	72,322	95,050	121,899
Growth rate	(%)	(9.0)	(7.6)	(10.5)	(9.0)	(5.6)	(6.4)
Composition	(%)	(30.8)	(29.5)	(29.3)	(28.4)	(22.9)	(20.6)
Oil		33,971	61,247	89,247	128,173	208,748	297,302
Growth rate	(%)	(10.2)	(10.32)	(7.8)	(7.5)	(10.2)	(9.2)
Composition	(%)	(56.7)	(63.3)	(55.6)	(50.4)	(50.2)	(50.2)
Natural gas				739	8,030	17,264	30,790
Growth rate	(%)	–	–	(–)	(61.1)	(16.5)	(15.6)
Composition	(%)			(0.5)	(3.2)	(4.2)	(5.2)
Hydro		826	1,147	2,184	2,248	2,265	2,265
Growth rate	(%)	(1.2)	(5.6)	(13.7)	(5.8)	(0.2)	(0.0)
Composition	(%)	(1.4)	(1.2)	(1.4)	(0.8)	(0.6)	(0.4)
Nuclear			1,670	16,129	40,619	83,885	122,507
Growth rate	(%)	–	(–)	(57.4)	(20.3)	(15.6)	(9.9)
Composition	(%)		(1.7)	(10.1)	(16.0)	(20.2)	(20.7)
Fire wood		6,706	4,165	3,448	1,874	1,804	1,786
Growth rate	(%)	(3.4)	(7.6)	(3.7)	(11.5)	(8.4)	(0.3)
Composition	(%)	(11.2)	(4.3)	(2.2)	(0.7)	(0.4)	(0.3)
Tidal				923	1,124	4,832	10,720
Growth rate	(%)	–	–	(–)	(0.4)	(33.8)	(22.1)
Composition	(%)			(0.5)	(0.4)	(1.2)	(1.8)
Solar						1,678	2,978
Growth rate	(%)	–	–	–	–	(–)	(15.4)
Composition	(%)					(0.4)	(0.5)
Wind							1,786
Growth rate	(%)	–	–	–		–	(–)
Composition	(%)						(0.3)
Electricity GWh		16,630	43,170	78,634	181,921	222,916	316,747
Growth rate	(%)	(18.4)	(17.2)	(12.7)	(10.9)	(11.1)	(9.2)
Total		59,963	96,828	160,547	254,890	415,517	592,032
Growth rate	(%)	(9.0)	(8.3)	(10.6)	(9.6)	(10.3)	(9.3)
Composition	(%)	(100.0)	(100.0)	(100.0)	(100.0)	(100.0)	(100.0)

Nuclear Engineers) have seconded staff to Bechtel to participate in the design.

This policy will be continued for the next two nuclear projects, 7 and 8, for which a site has been chosen at Gyema but the contracts have yet to be awarded. In the meantime KNE is training staff by technical co-operation agreements with Belgo-nucleaire and Motor Columbus in Switzerland. KNE will then be in a position to take over the leading role for the design engineering and management for projects 9 and 10, probably by working on a joint venture basis with an experienced architect engineer with the contracts being placed by Korea Electric with KNE. In this way KNE will gradually build up a complete competence for the design and construction of nuclear plant.

In a similar way Korean industry will develop expertise in the manufacture of heavy components by working under licence with established suppliers. Hyundai International now has an agreement with Combustion Engineering for the transfer of technology for the manufacture of nuclear steam supply system components and a plant for this is being built in Korea to Combustion Engineering design. Hyundai International also has an agreement with General Electric Corporation for the manufacture of turbines and generators. In addition Hyundai Shipbuilding and Heavy Industry has a ten-year licence agreement with Westinghouse for NSSS components and turbine generators. Dae Woo Heavy Industry has a joint venture agreement with Brown Boveri (Switzerland) for large turbine generators and a co-operation agreement with KWU (Germany). Several other companies have long term technology transfer agreements with US, European and Japanese companies for a large variety of mechanical power generation and electrical equipment.

These efforts are being complemented by a staff training programme. By 1986 it is estimated that some 7,000 engineers and 3,000 technicians will be required: 1,100 in manufacturing industry, 300 in construction, 2,000 in engineering, 3,600 for Korea Electric Co. for construction and operation, 2,100 research and development and 400 in regulatory and other fields. In addition to training centres in Korea at the Korea Atomic Energy Research Institute and Korea Electric Co. at Ko-Ri some 1,800 will be trained overseas under the various technology transfer and licence agreements between 1979 and 1986.

In this way Korea plans to establish a complete nuclear power capability by the late 1980s, early 1990s.

INDIA

The plight of India is made plain by reports in June 1979 that the worst power crisis in the history of India's most industrialised state, Maharashtra, brought almost all industrial activity to a standstill with nearly 1.5 million workers being laid off. In the area between Bombay and Poona all electricity to general industries was totally cut off. This was due to the delayed arrival of the monsoon; hydro-electric reservoirs ran dry. There is also a shortage of coal for the thermal power stations, and in any case a daily power shortfall of about 1,000 MW.[8]

India now has 580 MW of nuclear generating plant in operation in three power stations. Tarapur 1 and 2, each of 200 MW and RAPP-1 of 180 MW. The Tarapur stations use the BWR system and were supplied by General Electric and built by Bechtel as the Architect/Engineer. RAPP-1 in Rajasthan is a Candu type heavy water station supplied by Canadian General Electric with the turbine from English Electric Co. Ltd (Canada). Atomic Energy of Canada Ltd (AECL) with Montreal Engineering acted as Architect/Engineer and the Hindustan Construction Company took part in building the station.

Further nuclear development was continued under an agreement with AECL using the Candu design of reactor, but with Indian industry taking over the supply and construction of the stations. The first of these Indian stations, RAPP-2, a duplicate of RAPP-1, is to come into operating during 1980, again using an EEC turbine with AECL and Montreal Engineering as Architect/ Engineer. Four more stations are now under construction. Two of these each of 220 MW are being built at Madras, MAPP 1 and 2, by the Indian Department of Atomic Energy with the participation of Indian industry. The turbines are also being supplied by an Indian company, Bharat Heavy Engineering. These two stations are expected to come into operation in 1980 and 1982. Two more similar stations are under construction at Narora, NAPP 1 and 2, for service in 1983 and 1984. This will give a nuclear capacity of some 1,600 MW by 1985.

There are plans to increase the reactor unit size from 220 to 500 MW and to set in hand a programme to give 5,000 MW in operation by 1990, and 10,000 MW by the turn of the century, of which 3,000 is expected to be provided by fast reactors.[9]

In this programme the import component of the nuclear stations has been progressively reduced, from 45% for the first unit of Rajasthan, to 30% for the second unit, 12% for the Madras stations and is now at 9% for Narora.

A fast breeder test reactor, which is being built at Madras with French collaboration, is making steady progress. The long term aim of this programme is to use thorium as the fuel. The Indian resources of thorium, estimated at over 320,000 tonnes, are the largest in the world. The enriched fuel for the BWR stations at Tarapur is supplied from the US. Although India has a small reprocessing plant, built at the Trombay research centre the present plans are to store the LWR fuel after discharge in a special storage facility pending clarification of US policy on spent fuel reprocessing.

BRAZIL

The Brazilian Government has embarked upon a comprehensive nuclear programme that is intended to bring into operation some 10,000 MW of nuclear power by 1990. This will provide about 14% of the planned electricity supply. By the year 2000 the nuclear capacity will increase to 60,000 MW to give about 35% of the electricity supply from nuclear power.

The first nuclear station, due for completion towards the end of 1979 is ANGRA-1, a 626 MW PWR being built by Westinghouse. For future developments however the Brazilian Government has entered into a wide ranging agreement with Germany, covering all facets of the nuclear industry. Under this agreement a number of Brazilian–German companies are to be formed as joint ventures to provide the most effective means of transferring nuclear technology to the Brazilian industry. The key company on the Brazilian side is Empresas Nucleares Brasileiras SA NUCLEBRAS which is the Brazilian partner in (German interest in parentheses):

- NUCLEN – Nuclear Engineering, with Kraftwerk Union (25%)
- NUCLEP – Heavy components, with KWU, GHH and Voest-Alpine (25%)
- NUCLEI – Enrichment, with Steag and Interatom (10%)
- NUSTEP – Enrichment plant, construction and operation, with Steag (50%)
- NUCLAM – Uranium prospecting*, with Urangesellschaft

*Reserves of uranium ore equivalent to almost 200,000 tonnes of uranium have already been identified or estimated.

In addition NUCLEBRAS will, through technology transfer agreements, be responsible for fuel element fabrication using KWU technology; yellow cake production using Pechiney Ugine-Kuhlman technology; and reprocessing using technology from the German KEWA–UHDE group.

Under this programme eight 1,245 MW PWR stations will be built by NUCLEN based on the KWU design to give a capacity of 10,000 MW by 1990. The first of these stations, ANGRA-2 is planned for operation by 1985. All the civil work and 30% of the plant components will be supplied by Brazilian industry. NUCLEN will be the architect/engineer and responsible for the balance of plant: KWU will supply the NSSS and the turbine generator. ANGRA-3, now being planned is due for operation in 1986. In this programme the extent of Brazilian supply will be successively increased. NUCLEN will develop the basic design of the nuclear power plant and will contract with Brazilian companies for detailed design, plant construction and erection, and thus promote the development of a domestic industry, by placing orders, co-ordinating the transfer of technology to Brazilian suppliers and providing technical assistance. As NUCLEN's Brazilian staff gain competence they will gradually assume more senior positions within the company and the KWU participation will be phased down accordingly. Part of the equipment will come from the heavy components fabrication company, NUCLEP (Nuclebras Equipamentos Pesados SA) which will initially have the capacity to supply one set – pressure vessel, pressuriser, steam generators – per year but this will be expanded to two or three and eventually to five sets per year. Brazilian staff are being trained in Europe at the plant of the KWU–GHH–Voest-Alpine consortium.

The enrichment company will use the jet-nozzle process being developed by Professor Becker of the Karlsruhe research centre – KFK – with Steag. The enrichment demonstration plant with a capacity of 200 tonnes separative work/year to be built by NUCLEI (Nuclebras Eriquecimento Isotopico SA) will be the first of its kind using this process.

Fuel element design and fabrication is to be carried out by Nuclebras. Technical assistance and training is being provided by RBU, the KWU nuclear fuel company. This will cover all stages, powder production, pellets, assembly and structural components fabrication. A plant with a capacity of 100 ton/year is now under construction. Similar arrangements, this time with the French company Pechiney Ugine Kuhlman cover the construction of a uranium concentrate (a yellow cake) plant.

Another technical agreement with the German reprocessing company KEWA–UHDE, the reprocessing plant construction group, will cover "reprocessing of the spent fuel and waste treatment".

The high quality of equipment required for nuclear work will be ensured through Quality Assurance procedures, and techniques are being developed with the participation of American and German specialists. An Institute for Nuclear Quality, similar to the German TÜV organisation will be formed to act as an independent expert and inspection authority. Licensing of nuclear plant is the responsibility of the Comissao Nacional de Energia Nuclear which draws on technical assistance contracts with the US ERDA and the German TÜV.

To support the technology transfer programme Brazilian staff is being trained abroad in Germany and elsewhere at the facilities of the overseas partners, both in industrial and research installations. An agreement between the German and Brazilian Governments gives Nuclebras engineers and scientists access to the nuclear research centres at Julich, Karlsruhe and Geesthact. German experts from these centres can also be assigned to assist research and development projects in Brazil.

By the means of these agreements Brazil will in due course acquire a complete nuclear competence, as well as a substantial nuclear plant capacity and supporting industry. With electricity consumption growing at over 12% per year, the large hydro power potential will be fully utilised by the year 2000. Brazil then has a short 20 years to establish its nuclear industry.

NUCLEAR POWER IN THE CMEA COUNTRIES*

Another co-operative venture is between the CMEA countries which will lead to a massive expansion of nuclear power in those countries based upon Russian technology. A summit meeting of the prime ministers of the CMEA countries, decided, in June 1979, that a greatly increased nuclear power programme should be put in hand. This comes at a time when the Soviet Union, the principal energy supplier in the CMEA group, is experiencing a decline in the rate of increase of its oil production. The incentive is not only to overcome shortages of fossil fuels, but because nuclear power can generate electricity some 15–20% below the

* Council for Mutual Economic Assistance: USSR, German Democratic Republic, Bulgaria, Hungary, Czechoslovakia, Roumania, Yugoslavia, Poland and Cuba.

cost of fossil fuelled plant.* Antoli Alexandrov, President of the Soviet Academy of Sciences has declared that "without nuclear power countries such as the US, West Germany and Japan would quickly lose their economic importance".[10]

The new CMEA nuclear programme then calls for an installed capacity of 150 GW by 1990; a very large increase beyond the previous plan for which the target was 30 GW by 1980. But since the new programme has been decided upon with the knowledge that the 30 MW is unlikely to be fully achieved by 1980 (the present operating capacity is about 16 GW but with a number of stations due to come into operation within the next 1–2 years) it represents the renewed determination to devote a major industrial effort to the expansion of nuclear power capacity.

In this joint programme the USSR is the major partner, having developed a substantial nuclear capability over the past 30 years. The Russian nuclear programme has followed four main lines:

Pressure Tube Reactors
Pressurised Water Reactors
Fast Reactors
Heat Reactors

Pressure tube reactors

Russia can claim to have put into operation in 1954 the world's first nuclear power station, the 5 KW(e) AM-1 at Obinsk, a light water cooled, graphite moderated pressure tube reactor. This reactor type is a purely Russian development which has been built only in the USSR. It has some similarity to other pressure tube reactors which were developed later, such as the (now abandoned) UK design of the Steam Generating Heavy Water Reactor (SGHWR), but uses a graphite moderator instead of heavy water. The fuel is low enriched uranium. These reactors are of a modular construction, each channel or tube can be fuelled individually with the reactor on load. Such a design lends itself to the production of low-irradiated plutonium suitable for military use. The second reactor of this type to be built was indeed the 600 MW Troitsk station in 1958 which is described as being for military "research".

* Because of difficulties in establishing an equivalent currency exchange rate it is not easy to make a direct comparison of nuclear generating costs in Russia with those of other countries. Within Russia however it is said that the cost of power at large nuclear stations is on average 15–20% lower than at stations operating on fossil fuel. Actual figures quoted for 1975 are:

Novovoronezh NPP – 0.642 kopecs/kWh
Krivorozhskaya 3,000 MW coal plant – 0.895 kopecs/kWh
Konakovo 2,400 MW fuel oil and gas – 0.712 kopecs/kWh.

The modular construction also enables large reactor units to be built up and after gradually working up the civil design through the stages of 100 MW of Beloyarsk-1 of 1964 and 200 MW with Beloyarsk-2 of 1967 the standard model of a 1,000 MW reactor was reached — designated RBMK-1000 with two 500 MW turbines.

A whole series of these reactors has been built and will be brought into operation between 1973 and 1980:

Leningrad	1, 2, 3 and 4
Kursk	1, 2
Chernobyl	1, 2
Smolensk	1, 2

The output of these reactors has now been further increased and two 1,500 MW units are being built at Ignalina in Lithuania for operation in the early 1980s; two further stations are planned to give a total of 6,000 MW on one site. The 1,500 MW stations have two 750 MW turbines. It has been suggested that in the future, stations of up to 2,000–2,500 MW will be built with 1,000 MW turbines. It is worth noting that so far no pressure tube RBMK stations have been built outside the USSR.

Pressurised water reactors

The USSR also developed a pressurised water reactor. The first of these, the VVER-1, 280 MW station, started operation in 1964 at Novovoronezh. The output of this design was successively increased with Novovoronezh-2, 365 MW in 1969 leading to the Novovoronezh-3, of 440 MW in 1971. This 440 MW station was then adopted as the first standardised design and a series of further stations were built. Novovoronezh-4 in 1972; Kola-1, 1973; Kola-2, 1974; Oktemberjan-1 in Armenia, 1976; with a second station Oktemberjan-2 due to start up in 1980, Rovno-1 and 2 due to start in 1980 and Kola-3 and 4 again for start up in 1980. The VVER-440 station also became the standard export model with stations being built in Bulgaria, the German Democratic Republic, Czechoslovakia and Hungary; two stations of this design have been supplied to Finland, one is to be built in Turkey and one is proposed for Cuba.

The next stage in the development of the PWR was to increase the size to 1,000 MW(e) with two 500 MW turbines. Four of the 1,000 MW VVER stations are now being built and will shortly come into operation. Novovoronezh-5, 1979; Kalinin-1, 200 km north west of Moscow, 1980; Rowno-3 in the West Ukraine, 1980 and Nikolajew in the S. Ukraine, 1980. This last station will operate in conjunction with a pumped storage scheme and the warm water discharge will be used for horticulture development.

Further VVER-1000 stations are being planned for Kalinin-2, 3 and 4, Nikolajew-2, 3 and 4, Khmelnitskiy-1, 2, 3 and 4 in the West Ukraine which will also deliver power to Poland through a 750-kV grid connection.

Standardising on one design for a large continuing programme makes it possible to mass produce the main components on a production line system; this is now being done at the Atommasz engineering works at Volgodonsk, which will produce steam generators and reactor pressure vessels for the VVER reactors. This is a completely new plant, the first stage of which was completed in 19 7 and construction of a second stage is now underway to be in full operation by 1982. It will have the capacity to produce eight sets of reactor components (pressure vessels, steam generators and pressurisers) per year.

Fast reactors

The Soviet Union showed an early enthusiasm for fast breeder reactors and BR-5 at Obinsk, completed in 1959 and now used as a fuel and materials test reactor was one of the first sodium cooled fast reactors to come into operation.

The second fast reactor BOR-60, a 60 MW (thermal) reactor in operation at Dimitrovgrad in 1969, is also used for test purposes.

This early work led to the design and start-up in 1973 at Shevchenko of the first industrial fast reactor BN-350 with a thermal power of 1,000 MW. This reactor is equipped with a back pressure turbine, the steam from which is used in a desalination plant to produce 120,000 m^3/day of desalinated water from the Caspian Sea, whilst also generating 150 MW of electricity, instead of 350 MW from a fully condensing turbine. Despite some early troubles which provided enforced experience in repair procedures – cutting main sodium pipes and cleaning steam generators and circuits of sodium/water, sodium/air reaction products – proceeding to the next step was considered as justified. This is the BN-600 reactor, thermal power 1,470 MW now being built at Beloyarsk and due for completion in 1980. Whilst some of the components will be similar to BN-350 the six-loop design of the earlier reactor has been rejected in favour of an integral vessel. This is a similar design concept to the 250 MW PFR at Dounreay in the UK and to the French Phenix 250 MW and SuperPhenix of 1,200 MW now under construction at Creys Malville.*

* The BN-600 which started operation on 8 April 1980 is at present the world's largest operating fast reactor.

The Russians are now working on the design of a larger reactor, the BN-1600, which should have improved breeding characteristics, (a doubling time of less than 6 years), higher reliability and better economics, for industrial operation on a wide scale by the end of the 1980s.

Fast reactor fuel development has been based on oxide fuel, but other fuels such as carbide and nitride have also been studied.

While the main thrust of the Russian fast reactor work is on the sodium coolant, gas coolants, notably helium, and the dissociating nitrogen tetroxide, $N_2O_4 \leftrightharpoons 2NO_2$, used as a gaseous coolant in the reactor, condensing to liquid in the turbine, are also being considered.

Heat reactors

In addition to the use of nuclear energy for generating electricity the USSR has always shown an interest in using nuclear heat for chemical processing and particularly for district heating. Only about 25% of primary energy in the USSR is used for electricity production, the remainder is for industrial and domestic heating. A first step was made with the construction of four small light water graphite reactors of 12 MW each at Bilibin in Northern Siberia, which supplied heat and power to the local settlement.

Other small heat and power units have been designed based on the boiling water reactor and a small heat and power plant is being built 25 km from Odessa. Heat only reactors using organic coolants and producing hot water at 150°C with an output of 15 MW(th) are also being developed. It is planned that over the next ten years several hundred such nuclear heating plants will be built in major cities throughout Russia.

Fuel cycle

The USSR and the other CMEA countries could hold as much as 20–30% of the world uranium resources. Uranium ore mined in Czechoslovakia is processed to the concentrate stage before being shipped to Russia. Little is known of the Russian fuel cycle facilities beyond the fact that they must exist on a substantial scale. Russia is the sole supplier of enriched uranium for the whole of the CMEA nuclear programme, and also has an appreciable export to other countries, including W. Germany and France at a price marginally below that of the US Department of Energy. Russia also requires that the spent fuel from its exported VVER stations is returned for reprocessing, and is thus able to maintain a strict control over all fissile materials. No fuel cycle activities are carried out in any of the other CMEA countries.

CMEA co-operation

Although the fuel cycle remains under sole Russian control there is a growing co-operation within the CMEA countries on the construction of nuclear power stations and on the interchange of electricity. The nuclear power production is based on the VVER-440 reactors which are being built with Russian assistance in the CMEA countries. Although of Russian design, and with the main reactor components being supplied by Russia, the CMEA countries are being encouraged to build up their own expertise according to their industrial capacity, concentrating on particular items rather than trying to develop a complete nuclear manufacturing capability. By treating the CMEA programme as a single entity a large market is provided for those specialised items of equipment which could be manufactured by some of the less developed CMEA countries whose skills and experience would also be increased by participating in the overall programme under the guidance of the main partners. In the terms of the slogan coined by the Hungarian Power Plant Investment Co. (ERBE) they will "teach by working and work by teaching".

Of the CMEA countries, Czechoslovakia has gone furthest in developing its nuclear capacity and by 1981 will have the capacity to produce five sets per year of complete reactor equipment and primary circuit components for VVER-440 stations.

With the full development of nuclear manufacturing capability of all the CMEA countries it would then seem that the target of 150 GW of nuclear power by 1990 is not beyond reach. The earlier programme of 30 GW by 1980 was on the basis that 20 GW would be built in Russia and 10 in the other CMEA countries. If the same proportion is maintained the new programme would have 100 GW in Russia and 50 GW in the other CMEA countries. For Russia the achievement of this programme will be assisted by the policy of expanding existing sites to between 4,000 to 12,000 MW which should enable a rapid expansion to take place. Omitting the smaller early development stations the Russian programme could then be based on some eighteen different sites for which a substantial expansion has already been outlined as shown in Table 43.

For the longer term proposals have been made[11] to concentrate on nuclear power parks sited in remote areas. These would be vast complexes with 40–50,000 MW of power on one site, together

Table 43: *USSR nuclear power stations*

	No.	*MWe*	
RBMK, water, graphite reactors in operation or under construction:			
Leningrad	1–4	4,000	
Kursk	1 and 2	2,000	
Chernobyl	1 and 2	2,000	future expansion to 12,000
Smolensky	1 and 2	2,000	
Ignalina (Lithuania)	1 and 2	3,000	future expansion to 6,000
VVER, pressurised water reactors in operation or under construction:			
Novovoronezh	1–5	2,500	
Kola (Murmansk)	1–4	1,760	
Oktemberjan (Armenia)	1–2	880	
Rovno (W. Ukraine)	1–3	1,880	
Kalinin	1–4	4,000	
Nikolajew (S. Ukraine)	1–4	4,000	
Khmelnitskiy (W. Ukraine)	1–4	4,000	
Sites named for future stations:			
Aktash (Crimea)		6,000	
Konstantinovskaya		4,000	
Zaporozhye		4–5,000	
Balakovo			
Volgodonsk		4–5,000	
Saratov		4–5,000	

with all the fuel cycle facilities – fuel fabrication, reprocessing waste management and waste disposal. This has been put forward to avoid the constraints that would be met if all the nuclear power stations were built within the main area of demand in European Russia to the West of the Volga–Volga Baltic Canal in which 60% of the population of the USSR live and which accounts for a large part of the agricultural production and contains much arable land. There could then be difficulties in finding sufficient sites and adequate supplies of cooling water. The nuclear centres could be built in areas with a low density of population where there is enough land – unfit for other use and of low value – and where cooling water resources are available.

The positive advantages claimed for such nuclear centres are: that it would be cheaper to build a number of stations on one site; the transport costs for moving spent fuel to a reprocessing plant, and the waste to a disposal facility, would be avoided; the conditions of waste management would be simplified; the costs of safety and efficiency of safeguards would be reduced by a stricter centralised control and the minimising of external transport; and at a time when most of the power stations will be fast reactors, reprocessing and refabricating the fuel on one site would reduce out-of-reactor time and enable higher breeding gains to be obtained. These nuclear centres would not only generate electricity for transmission to the consuming areas 1,500–3,000 km distant, but would also be used for the large scale production of hydrogen as an energy carrier.*

For the other CMEA countries the nuclear programme is based mainly on the Russian pressurised water reactors with the exceptions of Roumania which recently came to an agreement with Canada covering the construction of four 600 MWe Candu heavy water, natural uranium reactors and Yugoslavia which is purchasing a 630 MW(e) PWR from Westinghouse.

Long term nuclear plans for some of these countries have been given for 1990 as:

Bulgaria	7,700 MW
Hungary	5,700 MW
Poland	8,000 MW
Czechoslovakia	12,000 MW
Roumania	6,000 MW

If East Germany and Yugoslavia are also included 50,000 MW should be attainable. The detailed figures for stations at present in operation, under construction, or planned gives a total of 27,620 MW (Table 44).

Even though this 150 GW by 1990 may seem an ambitious programme, in the light of delays in meeting the present target, the CMEA countries foresee an even greater expansion in the last 10 years of the century to give a total which has been put as between 250–450 GW by the year 2000.

THE FAST REACTOR

In recent years the fast reactor has become the particular bogey, or target, of the nuclear opponents. There are several reasons why this should be so. The word 'fast', which refers only to the velocity of the neutrons in the core of the reactor, can be misrepresentated to the public by analogy with a fast motor car, as implying that the reactor is more difficult to control and more

* A more recent article by the same authors, Dollezhal and Koryakin, reprinted in *Soviet News*, 13 November 1979, has been widely misrepresented as signs of a Russian environmental movement seeking to resist nuclear power development. But these are not negative proposals, they are a positive response to the very large expansion envisaged for nuclear power in the USSR.

Table 44: *Nuclear power stations for CMEA countries except USSR*

	In operation	*Under construction*	*Planned*
Bulgaria			
Kosloduj 1, 2	880		
Kosloduj 3, 4		880	
Cuba			440
Czechoslovakia			
Bohunice 1	110		
Bohunice 2 A and B		880	
Dukovany 1, 2		880	
Dukovany 3, 4			880
Leviece 1–4			3,520
Germany D.R.			
Nord 1, 2, 3	1,320		
Nord 4		440	
Magdeburg 1–4		1,760	
Magdeburg 5–8			1,760
Nord 5–8			1,760
Hungary			
Paks 1, 2 (Danube)		880	
Paks 3, 4			880
Poland			
Zarnowice 1 and 2		880	
Zarnowice 3			1,000
New site			4,000
Roumania			
Olt (Bucharest)		440	
Cernavoda (Danube) Candu			2,400
Yugoslavia			
Krsko (Slovakia) Westinghouse		630	
Vir			1,000
Totals	2,310	7,670	17,640

Source: GHG.

likely to cause serious accidents. This is of course not so. The other name often used is 'breeder' reactor – a reference to the fact that a fast reactor can be designed, by absorbing excess neutrons in fertile uranium-238 placed around the core, to produce more fissile plutonium than it consumes. This concept of 'breeding', of getting something for nothing, of producing more plutonium than is used seems to be disturbing and alarming. But as Dr Marshall has observed:[12]

> "The 'fast breeder reactor' does not, in fact, breed fast. It simply uses fast neutrons and breeds rather slowly while incinerating plutonium."

The fast reactor core is a plutonium burner, it is the uranium blanket surrounding the core in which new plutonium is formed, and from which it can be recovered, that determines whether the fast reactor produces a net excess of plutonium. This net excess is usually quite small, and less than the plutonium which is inevitably formed in the conventional thermal reactors. The fast reactor should then be preferred from a non-proliferation point of view since the plutonium is used and the total world production can then be limited, whereas with the use of thermal reactors on their own the world production of plutonium would rise indefinitely.

The principal importance of the fast reactor lies in its ability to convert a far greater proportion of the energy available in uranium than is the case for the thermal reactors. A system of fast reactors with an adequate breeding gain can provide nuclear power, for all practical purposes, free from any uranium supply constraints. On present knowledge the "Reasonably Assured" and "Estimated Additional" uranium resources, at a notional recovery cost of up to $130/kg of uranium (the market price could be much higher), are put at just under 5 million tons. Used in thermal reactors all of this could be committed before the end of the century. Even if the forecast of an extra 7–15 million tons of "Speculative" reserves can be found there could still be acute pressure on supplies of uranium at a reasonable price by 2025. However with a programme of fast reactors the available uranium – even leaving aside the speculative reserves – would be sufficient to maintain nuclear generation for many hundreds of years.

The fast reactor can also "burn up" the depleted uranium – the uranium-238 discharged from the enrichment plants after the greater part of the ^{235}U has been concentrated into the enriched uranium product. Used in this way in a fast reactor the present stocks of depleted uranium now stored in the UK from the operation of the small British enrichment plant would have an energy content equivalent to 40,000 million tons of coal. This is of the same order as the whole of the estimated coal reserves of the UK. The depleted uranium is, however, already processed and in store above ground available for immediate use whereas the equivalent amount of coal has yet to be mined with an inevitable and very high loss of life as well as causing considerable environmental disturbance.

The uranium recovered by reprocessing spent fuel could provide another vast source of energy. In the UK the annual arisings of reactor depleted uranium magnox fuel (about 780 tonnes/year) would, if fissioned in a fast reactor, equate to 1,600 million tonnes of coal or 6,250 million barrels of oil. This is almost twice the present annual oil production of Saudi Arabia. It would however require the operation of something over 500 GW(e) of fast reactors to produce this energy in one year, or 10 GW over 50 years.

From the point of view of reactor safety, far from being at a disadvantage, the fast reactor could, in some respects, offer a considerable advance over the conventional thermal reactor systems. Recent tests in the UK with the 250 MW PFR at Dounreay have shown that the pool-type sodium cooled reactor has an intrinsic safety feature in that the reactor could survive a total loss of electrical supplies for an extended period. If the primary pumps are tripped, and their auxiliary motors cannot be started they would come to rest in 2000 seconds, but natural circulation can take over so that with the reactor shut down the decay heat can be removed from the core and safe temperatures can be maintained indefinitely. Even if the decay heat rejection systems were to fail the large thermal capacity of 1,000 tonnes of sodium in the primary pool is sufficient to delay boiling of the primary coolant for at least 25 hours.

With its more efficient use of natural uranium – where the fast reactor can extract up to sixty times more heat from a given quantity of uranium than a Light Water Reactor – the fast reactor can also show, as a direct consequence, considerable environmental and radiological advantages. The main contribution to the collective radiation dose from nuclear power comes from the natural radioactivity of the mill tailings waste at the uranium mine. Since it uses less uranium, in comparison with a LWR, the mining and milling operations for a fast reactor fuel cycle are reduced by at least a factor of 50. The radiological impact of normal reactor operations on the population and the environmental disturbance of mining would then be correspondingly lower.

Although the capital costs of the first commercial scale Fast Reactors are expected to be higher than those of the present LWRs by a factor of 1.5–2 these costs will fall with increasing experience, the scaling-up of manufacturing capacity and with an increasing number of plants. However any extra capital cost may be more than made good by the fuel cost savings. A further consideration is that the wider use of fast reactors would relieve the pressure on uranium requirements and probably slow down increases in uranium price: this would then be reflected in lower fuel cycle costs for the thermal reactors as well.

The costs of developing a breeder reactor and its fuel cycle are high. These costs are then only likely to be borne by countries with large energy requirements, but without indigenous uranium resources. The main fast reactor development has then been in the countries of W. Europe and Japan, but the USA and especially the USSR have also invested a large effort in this reactor system. There are at present three demonstration fast reactors in operation: BN-350 in USSR, Phenix in France and PFR in the UK, each of 250 MW. There are also experimental fast reactors in USA, USSR, France, UK, Japan and Germany. A first commercial scale reactor, the 1,200 MW(e) Super Phenix, is now nearing completion in France and the Russian BN-600 has now come into operation. Over the next twenty years there will be a gradual penetration of the fast reactor for nuclear energy production. It is possible that 50,000 MW of fast reactor capacity could be in operation by the year 2000.

Most of the present development has been concentrated on the liquid metal cooled reactor. Some work is also being done on the alternative of a gas-cooled fast reactor.

REFERENCES

1. Notiziario, Comitato Nazionale per l'Energia Nucleare, Rome, August 1979.
2. Financial Aspects of Nuclear Power Programmes, CN-36/7, IAEA, Salzburg Congress, May 1977.
3. AIF-Danatom Conference on Financing Nuclear Power, Copenhagen, September 1979.
4. R.K. Lester, *Nuclear Power Plant Lead Times*, The Rockefeller Foundation/The Royal Institute of International Affairs, November 1978.
5. *World Development Report*, The World Bank, 1979.
6. Mumir Khan, *Nuclear Energy and International Cooperation: A Third World Perception of the Erosion of Confidence*, The Rockefeller Foundation/The Royal Institute of International Affairs, September 1979.
7. Choi, Chang Tong, Korea Electric Co., "Nuclear Power Programme and Localisation in Korea", paper given at European Nuclear Society Conference, Hamburg, May 1979.
8. *The Guardian*, 18 June 1979.
9. H.N. Sethna, *IAEA Bulletin*, 21 (No. 5), October 1979.
10. *Financial Times*, 19 June 1979.
11. Dollezhal, Koryakin *et al.*, Paper No. IAEA-CN-36/334, Nuclear Energy Centres: Economic and Environmental Problems, IAEA Salzburg Congress, May 1977.
12. W. Marshall, Nuclear Power and the Proliferation Issue, Graham Young Memorial Lecture, University of Glasgow, February 1978.

6
How much nuclear plant in operation by the year 2000?

In attempting to estimate the amount of nuclear power that might be in operation by the year 2000, there are four questions that need to be answered:

How much is required?
How much could be used?
How much could be built?
How much can be expected to be built?

In an ideal world the answers to these questions should point to the same figure. In practice there are not only considerable differences between the four sets of figures but wide uncertainties within each set.

HOW MUCH IS REQUIRED?

From the different energy forecasts given in Chapter 2 the requirement for nuclear power can be expressed in terms of installed electric capacity in GW(e) by the year 2000. Table 45 also includes estimates of nuclear capacity made by the International Fuel Cycle Evaluation (INFCE) and the Kernforschungsanlage (KFA) Jülich in Germany. The INFCE study, carried out over two years, ended with a final Plenary Conference in February 1980. Sixty-six countries and five international organisations took part. The projection of possible nuclear power capacity are illustrated by a high and a low case. These are not limiting bounds but representative projections and it is emphasised that both higher and lower capacities are possible depending on decisions and events in the future. The study by KFA Jülich was made for the Rockefeller Foundation and the Royal Institute of International Affairs.

In isolating these figures it must still be remembered that they all form part of energy scenarios in which the supply of oil is assumed to increase by about 1.5 times the present figure and in which the output of coal is also assumed to go up by a factor of

Table 45: *Nuclear capacity in year 2000 (GW(e))*

	GW(e)
World	
WEC (Conservation Commission)	1,400–1,540
WEC (Cavendish) L4	1,300
WEC (Cavendish) H5	1,900
Interfutures	1,300
IIASA (1978) Low	1,500
IIASA (1978) High	1,850
INFCE	1,000–1,600
Jülich	900
WOCA	
UNICE	900–1,500
WAES	1,000–1,400
INFCE	850–1,200
Jülich	690
OECD	
Interfutures	990
WEC (CC)	800
IEA Steam Coal	650–830
INFCE	670–1,000
Jülich	610
Centrally planned	
WEC (CC)	560
Interfutures	210
UNICE	600
INFCE	250–450
Jülich	200

about 2.

The first of these assumptions must now be regarded as highly optimistic, the second will be difficult to achieve but could be considered as within grasp provided a sufficient effort is made. All the scenarios also assume extensive conservation measures; as indicated by the fall in the GNP/energy ratio, so little further can be expected from this direction.

The nuclear estimates of Table 45 ought then to be considered as minimum requirements and as the realisation of energy shortage strikes home there could be strong pressure to raise the figures if this were possible. But whether this can or will be done rests on the answers to the three following questions.

HOW MUCH COULD BE USED?

There is an upper limit to the nuclear contribution to world energy supplies which is set by the fact that until the end of this century nuclear power will in the main be used for the generation of electricity. The estimates of nuclear power capacity can then be considered in the light of the possible electricity demand and the share that will be taken by nuclear power.

Since its introduction into general use some 50 years ago electricity consumption has grown at a faster rate than for energy demand in general, and for any particular fuel. World electrical generating capacity with a growth rate of 8% per year since 1955 has shown a doubling period of nearly 9 years (Figures 28 and 29).

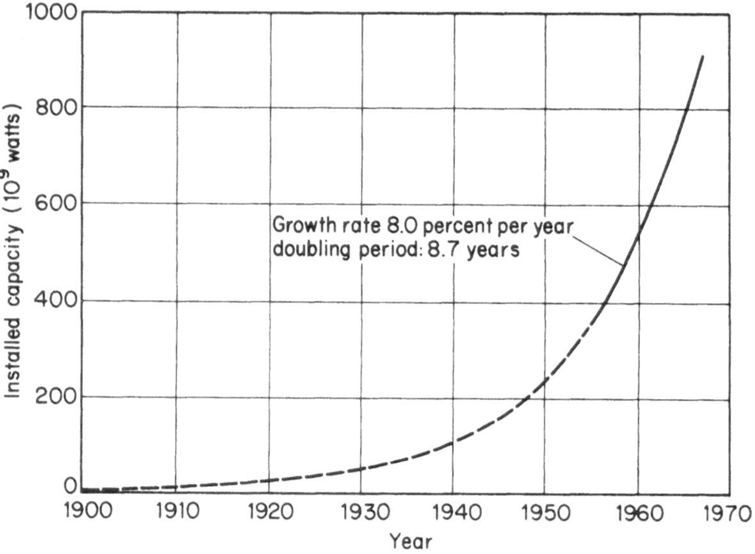

Figure 28: *World electrical generating capacity*

With its convenience, immediate availability and cleanliness at the point of use, electricity is the generally preferred carrier for energy use. While the rate of growth may be slowing in industrial countries, it still remains well above energy demand in those countries. In the USA for instance electricity growth was 6.85%/year over the period 1960–73. This is about 1½ times the growth of total energy demand and it was recently projected to maintain 4.7% per year for the rest of the century. On this basis the electricity fraction of equivalent primary fuel input in the USA could grow from the present figure of 30% to between 46–57% in the year 2000, with 50% being taken as a planning target.[1] In the UK electricity growth rates have now dropped in recent years to about 2% per year but this is after an expansion by a factor of 60 over a period of 55 years. But the fastest rates of growth are now being experienced in those developing countries

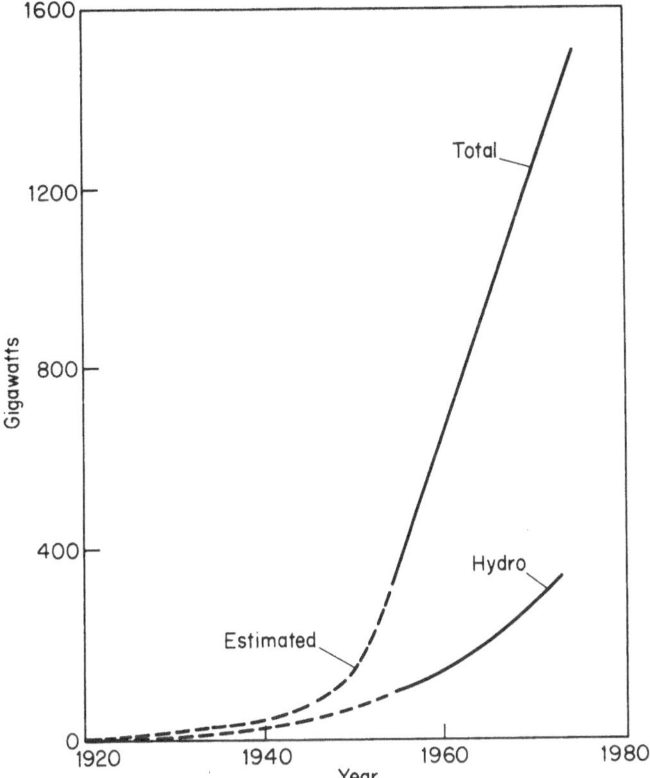

Figure 29: *Comparison of worldwide total electric with hydroelectric generating capacity. (Source: Ref. 2)*

which are moving into the industrialised phase. In India electric power generation has been growing by an average of 11.5% per year since 1950 and is expected to continue at a rate of 10% per year for the next thirty years. In Korea power demand growth over the period 1961 to 1976 has been at the astonishingly high rate of 18.3% per year. Although this rate will decline in the future it is still expected to average some 12% per year to the year 2000. By that time electricity would correspond to some 53% of primary fuel usage.

Brazil is another rapidly developing country which again has historically shown high and increasing rates of electricity growth averaging 11% per year over the past ten years. By the year 2000 electricity capacity is expected to grow to between seven to eight times its present value corresponding to average growth rates of 14–16% per year.

Compared with these figures the World Energy Conference in

their report *World Energy* adopt more modest growth rates for their forecast of world electrical demand (Table 46).[2]

Table 46: *Estimated potential world electrical demand (exajoules electrical output)*

World regional groupings	*1972*	*2000*	*2020*	*Average growth rate 1972–2020*
OECD countries	14.1	40	77	3.6%
Centrally planned economies	4.7	28	64	5.6%
Developing countries	1.7	9	32	6.3%
Total	20.5	77	173	4.5%

Source: Ref. 2.

By the year 2000 the assumed 77 EJ* for the world electrical demand would correspond to some 230 EJ of primary energy, about 40% of primary energy demand using the Alternative Demand Scenario. The present percentage of primary energy transformed into electricity is about 30% for OECD. By the year 2000 for OECD and the Comecon countries the figure is put at 50%. The developing countries are starting from a lower level and despite a more rapid rate of growth the share of electricity only reaches 18%. By 2020 world electricity production is expected to correspond to about half of the primary energy output.

The next question is the fraction of electricity demand that can be expected to be met by nuclear power. As nuclear capacity increases, its share of electricity production will rise. In some countries nuclear power is already making a substantial contribution to electricity supply. In Europe in 1977 the high figures include 22% in Sweden and Belgium and 17% in Switzerland. The Belgian figure will shortly rise to 40% with new nuclear stations due to come into operation in 1981/82. In the US the national figure of 11% masks wide differences between different states. In some states nuclear power accounts for over half the electricity supply, these are Vermont 79%, Main 65%, Connecticut 53% and Nebraska 50%, while at the other end of the scale the value for Ohio and Colorado is only about 1%. It can also be noted that in 1977 the US production of nuclear power at 11.8% of the total electricity, exceeded hydro power production at 10.4% for the first time.

* EJ = exajoules = 10^{18}J.

The World Energy Conference (Conservation Commission) believes that by 2000 the fraction of electrical energy supplied by nuclear power will tend towards 45% (corresponding to 18% of primary energy) rising to 65% by 2020 on a world wide basis. Combining this figure with their estimated electricity demand (Table 46) gives a potential world nuclear power installation of between 1,300–1,900 GW in the year 2000 and between 3,200–5,500 MW by 2020 according to the different assumptions for economic growth and energy demand. The projected values are shown in Table 47.[2]

Table 47: *Projected world nuclear power installation (GW(e))*

World regional groupings	*1975*	*2000*	*2020*
OECD countries	68	800	2,225
Centrally planned economies	7	560	1,850
Developing nations	1	180	925
Total	76	1,540	5,000

Source: Ref. 2.

On the other hand a table collated by the IAEA based on data from the International Nuclear Fuel Cycle Evaluation programme suggests that the share of nuclear power will only be between 20–27% of electricity capacity by the year 2000 but even on this lower figure the world nuclear capacity is put at 1,000–1,600 GW (Table 48). In terms of electricity production, the nuclear percentage is however higher, 26–35%, because of the higher utilisation factor of 65% assumed for nuclear stations compared with 48% for the average electrical system (Table 49). (It would correspond to 12–16% of primary energy.)

This is still strikingly lower than the World Energy Conference figure of a 45% share for nuclear power. It is difficult to assess which is likely to prove more accurate.

In some countries the percentage share of electricity taken by nuclear power can be expected to be pushed upwards. France plans to reach the 50% figure by 1985 and 70% by 1990. In the UK the Under Secretary of State for Energy has indicated that 50% of electricity generation from nuclear power should be the target for the year 2000. Ultimately, the Department of Energy has estimated that 80–90% of public electricity supply will be generated by nuclear power.[3]

With the use of oil and gas firing for electricity plant being discouraged by scarcity and high price and with the share of hydro

Table 48: *Estimates of the development of installed electrical and nuclear[a] capacity by main country groups (1000 MW(e))*

	2000	
	Total electric	*Nuclear (%)*
OECD North America	1,400–1,600	300– 450 (21–28)
OECD Europe	1,000–1,200	270– 400 (27–33)
OECD Pacific	400– 500	100– 150 (25–30)
Centrally Planned Economy	1,200–1,400	250– 450 (21–32)
Asia	640– 750	60– 75 (9–10)
Latin America	350– 450	40– 100 (11–22)
Africa and Middle East	240– 300	10– 25 (20–27)
World total	5,230–6,200	1,030–1,650 (20–27)
Industrialised countries	4,050–4,760	925–1,460 (23–31)
Developing countries	1,180–1,440	105– 190 (9–13)

[a]Based on the data obtained by the International Nuclear Fuel Cycle Evaluation Prog.

Table 49: *Estimates of total world electric and nuclear energy consumption (TWh)*

	2000	
	Total electric	*Nuclear (%)*
OECD North America	5,900–6,700	1,700–2,500 (29–37)
OECD Europe	4,200–5,000	1,500–2,200 (36–44)
OECD Pacific	1,700–2,100	570– 850 (34–40)
Centrally Planned Economy	5,100–5,900	1,400–2,500 (27–42)
Asia	2,700–3,100	300– 400 (11–13)
Latin America	1,500–1,900	200– 550 (13–29)
Africa and Middle East	1,000–1,300	50– 140 (5–11)
World total	22,100–26,000	5,720–9,140 (26–35)
Industrialised countries	17,100–20,000	5,120–8,140 (30–41)
Developing countries	5,000–6,000	600–1,000 (12–17)

power likely to fall from about 25% of world electricity supply at present to around 16.5% by the year 2000, the principal remaining fuel is coal. But coal will be increasingly required as the raw material for synthetic oil production: a larger nuclear share of electricity generation would then release more coal for this premium use. If the WEC figures for nuclear capacity in 2000 are increased on the assumption that the nuclear power share can be pushed even higher to 50% of electricity supply instead of 45% the installed capacity required would be 1,450–2,100 GW. These figures are then the upper limit of the amount of nuclear power that could be utilised. They correspond to about 20% of world primary energy supply in the year 2000.

A lower limit can be found in figures put forward in a study by the Jülich Nuclear Laboratory in Germany which gives the nuclear capacity as shown in Table 50.

Table 50: *Nuclear capacity (GW)*

	1995	*2000*	*2020*
World	550	900	2,500
OECD	400	600	1,400
Developing countries		80	380

These figures are based on the assumptions that:

(1) Electricity growth rates will be roughtly 3% for OECD, 4.5% for COMECON and 6% for developing countries.
(2) No country will go nuclear unless its grid size is about 4 GW.
(3) No nuclear plant will be in operation in 1995 unless it has been ordered or firmly planned, with site selection by now.
(4) For the subsequent period nuclear power plants will supply one-third, one-half or two-thirds of all new electricity depending on whether the country has high, intermediate or low resources in fossil fuel.

The first of these assumptions takes electricity growth rates that are lower than have been experienced in the last few years, even with the post-1974 recession. It also ignores the possible substitution of oil by electricity promoted by higher oil prices and lack of supplies.

The second already has a few exceptions, e.g. Pakistan. The development and availability of nuclear stations smaller than 600 MW(e) – perhaps in the range 200–400 MW(e) – would enable the electricity grid size to be correspondingly reduced and the electricity demand of some of the megalopolis now being created in developing countries could then justify nuclear plant.

The third assumption of a 15-year initial delay is the main influence on the low figure for nuclear capacity by 2000. As discussed in the next section this could be a realistic estimate of the time taken to bring new plant into operation. Once the initial delay has been overcome the rapid expansion of nuclear power, a near threefold increase in the 20 years 2000–2020 is in line with the other estimates. From this point of view the Jülich figures represent an awareness of the difficulty in starting large programmes rather than a doubt over the future expansion expected for nuclear power. But this more cautious estimate is useful in setting a lower limit for the period taken here; the nuclear capacity by 2000 could then lie within the wide range 900–2,000 GWe.

HOW MUCH COULD BE BUILT?

Is the upper limit of 2,000 GW by the year 2000 feasible? The answer depends on two factors, the time required to plan, authorise and construct new plant and the manufacturing capacity available to build it.

Time could be a major constraint, in particular the time that it takes to find and approve a site and to comply with the regulatory and licensing procedures. This will of course vary from country to country, and could be shortened if the perceived need is great, and there is a general wish for the new plant to be built more quickly, but at present the total time taken to bring a new nuclear station into operation could be of the order of ten years, of which perhaps just over half, about six years, would be taken by the actual building of the plant. As we have seen the Jülich study would put the total time at 15 years. A figure of 12–14 years is now common in the USA. These long delays are however not unique for nuclear power; 8–12 years seems to be required for almost any new energy project, as shown in Figure 30.

This means that even with a lead time of ten years new nuclear programmes even started today can only begin to make a contribution to world energy supplies by 1990. This gives a mere ten years before the admittedly artificial cut off date of the year 2000 for the period under review. It must then be assumed that it is only the plants being planned and ordered in the ten year period 1980–1990 that will come into operation in the next ten year period 1990–2000.

On this time scale it is unlikely that there will be any major advance in nuclear technology. There could, with large programmes, be a greater incentive towards standardisation of design, in particular for the specification and manufacture of key components. The stations to be ordered will continue the present pattern and will in the main be Light Water Reactors of 900–1,300 MW(e) output, with a smaller number of Heavy Water Reactors, of about 600 MW, and some gas cooled reactors. A limited number of fast

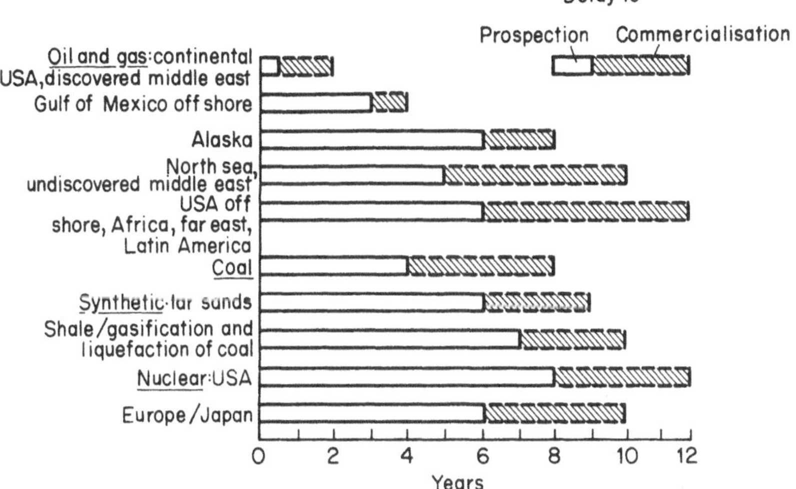

Figure 30: *Time taken to bring new power plants into operation*

breeder reactors may also be built, in a few countries, but these will be regarded more as first commercial prototype stations rather than the essential nuclear "workhorses".*

One major problem area will be in finding and approving new sites particularly in view of the increased sensitivity of the public as a result of the very wide, unfavourable publicity given to the Three Mile Island accident in the USA in 1979. A typical reaction to this accident has been summed up, from the result of a public opinion poll taken in Japan, as "Nuclear Power OK, but not in my town". In many cases it will be possible to consider placing the new stations as expansion of existing nuclear sites. Provided a sufficiency of cooling water is available there could be considerable advantages in this. There are already many examples of multiple station nuclear sites. The existing infrastructure, transmission lines to load centres, availability of experienced staff and usually a willing acceptance of nuclear power by the public in the immediate vicinity of an existing station could enable the planning, licensing and construction times to be appreciably reduced from the ten years assumed. As a rough guess one-third of the number of existing stations might be used in this way. Since there are some 500 different sites for stations in operation or under construction this could enable a rapid start to be made

* The INFCE study (Working Group 1A/2A) puts the power output from fast reactors at 20–38 GW(e) by the year 2000.

on about 150 GW which could come into operation in the latter part of the 1980s.

This expansion on a selected number of key sites has been taken furthest in the USSR, where the programme now planned envisages at least 4,000–5,000 MW on most of the present sites, rising in some cases to up to 12,000 MW on one site as for Chernobyl in the Ukraine. The possibility has been put forward of taking this policy to the limit with 40–50,000 MW on one site which would be complete with fuel manufacturing and reprocessing facilities. These would be remote sites with transmission lines of up to 1,500 km to the centres of population.[5] On the other hand the Russians are also developing the combined generation of heat and power which would require some stations to be built relatively close to the industrial or domestic users of the nuclear heat. Similar schemes for combined heat and power are being considered in both Sweden and Switzerland and other countries.

The other factor which determines the amount of plant that could be built is the available manufacturing capacity. This is very difficult to estimate because of the large number of companies involved, their diverse size and structure. Given sufficient time and forward planning almost any foreseeable programme could be met. Indeed it would be easier for a sub-contractor to meet a large programme of nuclear components built to a standard design by constructing a special nuclear manufacturing facility, rather than trying to produce a smaller number as a special item within his conventional manufacturing programme. The larger stations now being built in sizes of 900–1,300 MW would not take proportionately longer to manufacture and build than the earlier stations of 300–600 MW so that the total capacity could expand at a faster rate than in the 1960s.

It is however instructive to look at both past and present performance. For the United States market in the peak nuclear year of 1973 orders were placed for 40.965 GW, shared between four reactor suppliers. Of these Combustion Engineering took 15.8 GW and Westinghouse 13.4 GW. For the world market the year of peak orders was, 1974 when 65 GW were ordered from ten NSSS suppliers. The lead was taken by General Electric and Westinghouse with 14 GW and 10 GW while at the lower end were Atomic Energy of Canada and Asea-Atom of Sweden with 440 MW and 600 MW respectively. There are now some twenty NSS suppliers in the WOCA in ten countries. It might seem reasonable to assume that in a period of peak expansion these could accept, on average, between 5–10 GW per year of new orders, giving a total that could be ordered in the ten year period of about 1,600

GW.

Some guidance on what is actually being achieved with a sustained, planned construction programme can be obtained from France. Over the ten year period 1977–87 the French group Framatome/Creusot Loire, backed by the French manufacturing industry and with the support of the Government utility, Electricité de France, will bring into operation 40 Pressurised Water Reactor nuclear stations to give a total capacity of some 40 GW(e). At the same time the group has obtained at least four export orders. This rate of construction, which is now running at the rate of 5 GW/year has been achieved with a high degree of standardisation made possible by the firm commitment of the French Government to the programme. This has enabled the manufacturing industry to make long term plans and to invest in new plant and equipment so that the main components, pressure vessels, pressurisers, steam generators, pumps etc. can be produced on a production line basis.

Given the same political will and commitment to a standard design and a continuing production programme there are in the WOCA (world outside communist areas) countries the equivalent of some fifteen comparable reactor supplier groups who could probably do the same. These are in the industrial countries where the support of established component manufacturers already exists. Countries such as Spain, Switzerland, Benelux which have substantial heavy component suppliers are also included since they would be capable, working with an architect engineer to construct their own stations under licence. The list of countries is then:*

France	Spain	Benelux
Germany	Sweden	Japan (3 groups)
Italy	Switzerland	Canada
UK		USA (4 groups)

In addition there are countries such as Brazil and Korea who not only have plans for large nuclear programmes, but intend, progressively, to build up their own manufacturing competence: initially working under licence, but gradually taking over responsibility for the manufacture of components and the construction of plant. India should also be included with this group.

This then suggests that in round figures some 100 GW per year of nuclear plant could be brought into operation over the period

* Austria and Finland also have groups of industrial countries which have developed a considerable expertise in nuclear manufacture and supply.

1990–2000, giving a total of 1,000 GW of new nuclear capacity. Taking the mean of the historical ordering rate (1,600 GW) and potential capacity (1,000 GW) suggests that 1,300 MW of new plant could be built in the WOCA countries by the year 2000.

There are in addition the Eastern Bloc, CMEA countries. These will continue to use the standard Russian PWR design, but the size is being increased from 440 to 1,000 MW. With the co-ordination of manufacturing capacity in Czechoslovakia, Hungary, Poland, East Germany and with the support of the USSR, substantial expansion is planned with a common programme. Russia will also continue to build their water-cooled graphite moderated pressure tube reactors with the output increased to 1,500 or 2,000 MW. The total CMEA programme is planned to give 150 GW by 1990. By the year 2000 this could reach between 250–450 GW (say 300 GW) even assuming some delay in reaching the 1990 target. The new reactor component assembly plant "Atommash" which has been built in Russia at Volgograd on the Volga-Don canal will have a capacity to deliver eight complete reactor assemblies per year. The plant which was built in four years at a cost said to be more than £750 million has now started production and will reach full capacity by 1982, the time to fabricate a reactor pressure vessel being some three years.

Taking into account the 400 GW of plant now in operation or under construction, the total world nuclear capacity available by the year 2000 could then be 2,000 GW (400 + 1,300 + 300).*

This is, of course, a very crude assessment. It could be too high, since there is as yet no indication that governments regard the energy situation as sufficiently serious to set in hand the large and continued expansion of nuclear power required.

On the other hand the figure could be low. While the start up of these programmes may be delayed by a few years their acceptance will imply a firm commitment to nuclear power with expanding programmes being carried on until well into the next century. In this case there will be a very large investment in nuclear construction capacity which will be growing rapidly in the 1990s. The amount of new plant brought into operation in the second half of the 1990s could then rise well above the average figure assumed. From this point of view any attempt to freeze the picture to meet the year 2000 time frame will give a distorted view. In a period of rapid growth one or two years additional capacity would make a large difference to the total figure.

From this analysis it can be suggested that provided an early

* 400 GW existing capacity; 1,300 GW estimated for WOCA; 300 GW estimated for CMEA.

start is made on establishing firm programmes the manufacturing capacity need not be a limiting factor on the amount of nuclear plant that could be in operation by the year 2000. The 2,000 GW assumed is however uncomfortably close to the upper figures of the estimates in Table 47, and any delays in starting major world expansion would restrict the option of nuclear power to the lower end of the range.

It is then necessary to make the final assessment of how much nuclear power can be expected to be built by the year 2000.

HOW MUCH NUCLEAR POWER CAN BE EXPECTED TO BE BUILT BY THE YEAR 2000?

The figures of operating nuclear capacity show that nuclear power is already established in some countries and states as a substantial source of electricity. Yet despite the need for a new energy source, and despite the evidence that nuclear power has demonstrated that it can fulfil this need, there has been a marked falling off in recent years of new orders for nuclear stations. This is shown in Figure 31 which plots the annual ordering rate for nuclear reactors.

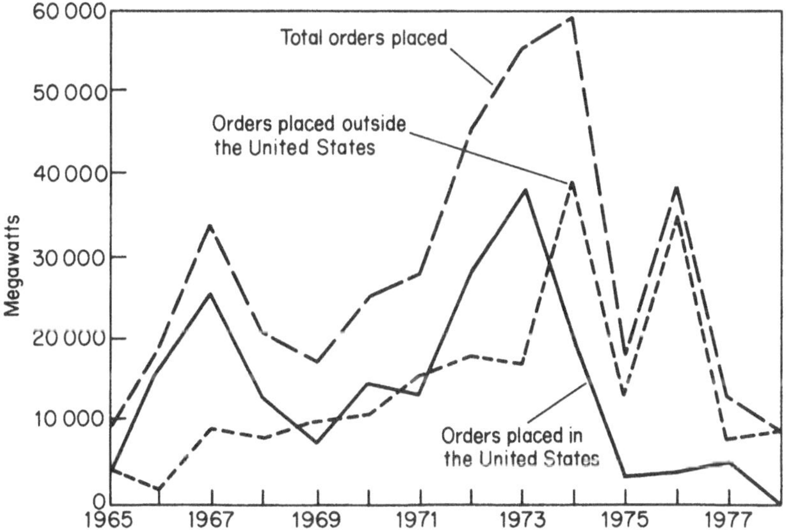

Figure 31: *Annual ordering rate for nuclear reactors. (Source: Kidder Peabody & Co.)*

This fall in nuclear orders can be ascribed to two factors which may even be inter-related: the continuing opposition of part of the public to nuclear power and the recession which followed the oil price rise of 1973/74.* The dramatic fall from the peak year of 1974 suggests that it is the latter which is the dominant factor. The opposition to nuclear power has grown slowly since the mid-1960s and while it could be expected to reduce the rate of expansion it could hardly bring about such a sudden reversal over the short period of one year.

The effects of the recession as indicated by a slowdown or even a fall in the rate of growth of GNP was most marked in the industrialised countries. Table 51 shows the percentage change over the years 1974–78.

Table 51: *Percentage change in GNP*

	1974	*1975*	*1976*	*1977*	*1978*[a]
OECD	5.1	−1.0	5.2	3.6	3.75
USA	−1.4	−1.0	6.0	4.9	3.75
EEC	1.7	−1.6	4.7	2.2	2.60
Japan	−1.2	2.3	6.0	5.0	5.75

[a]Provisional statistics.

This has been paralleled by a fall in the rate of increase in energy consumption (Table 52).

Table 52: *Energy consumption of primary sources (MTOE). (Combustible solids and liquids, natural gas, hydro/geo/nuclear power.)*

	1973	*1976*	*1977*	*1978*[a]	*Percentage change*		
					1976/73	*1977/76*	*1978/77*
OECD	3,528	3,526	3,568	3,657	−0.1	1.2	2.5
USA	1,772	1,780	1,812	1,854	0.2	1.8	2.3
EEC	989	961	964	985	−1.0	0.3	2.2
Japan	351	348	350	362	−0.3	0.6	3.4

[a]Provisional statistics.

It is only by 1977 or 1978 that energy consumption regained the pre-recession figure of 1973 (except for the USA where there

* Professor Tubiana: (Le Refus du Réel, Editions R. Laffont, Paris 1977) has suggested that the psychological stress following the oil price induced recession has contributed to the growth of the anti-nuclear movement.

has been a small but continuing increase).

The effect on electricity demand was less marked, but although production continued to grow it was at a slower rate than before (Table 53).

Table 53: *Electricity production of OECD (GWh)*

1970	*1971*	*1972*	*1973*	*1974*	*1975*	*1976*
3,518,070	3,706,529	4,010,528	4,306,224	4,408,164	4,456,841	4,757,169

This slowdown in electricity demand had a marked effect on the building programmes of the electricity supply companies. These programmes are planned some 8–10 years in advance, and since in the late 1960s/early 1970s demand was rising steadily new construction for the mid-1970s was planned to meet a substantial increase. When this did not materialise there was a postponement or cancellation of new power station projects, and since nuclear stations represent a higher proportion of new plant orders this has reduced the fraction of nuclear plant in the total supply system for the subsequent years. The cancellations of orders have also fallen disproportionately upon nuclear stations since they are usually of larger output, more capital intensive and take a longer time to build. The contributory effect of high interest rates and inflation in slowing down the US programme has already been noted (pp. 75–77).

This has led to a decline in the rate of penetration of nuclear power in electricity supply, as shown in the fall off from the straight line substitution expected from a Fisher Pry analysis (Figure 32). In this the fraction of the market taken by a new product, or process, F, expressed as the fraction $\log F/(1 - F)$ plotted against time, follows a straight line law, over long periods of time and for almost complete substitution of one product by another.

When such a curve is plotted for the fraction of the total electricity supply met by nuclear power, taking 1965 as the year of introduction of fully commercial nuclear power plants, the twelve year period up to 1977 can be seen to be a period of orderly and predictable growth. But the last few years have shown a marked falling away from the past trend (Figure 32).

There is then a seeming paradox in that the increase of oil price which gives nuclear power an even greater economic advantage over conventional oil-fired electricity generation and might be expected to lead to an expansion of nuclear programmes has in

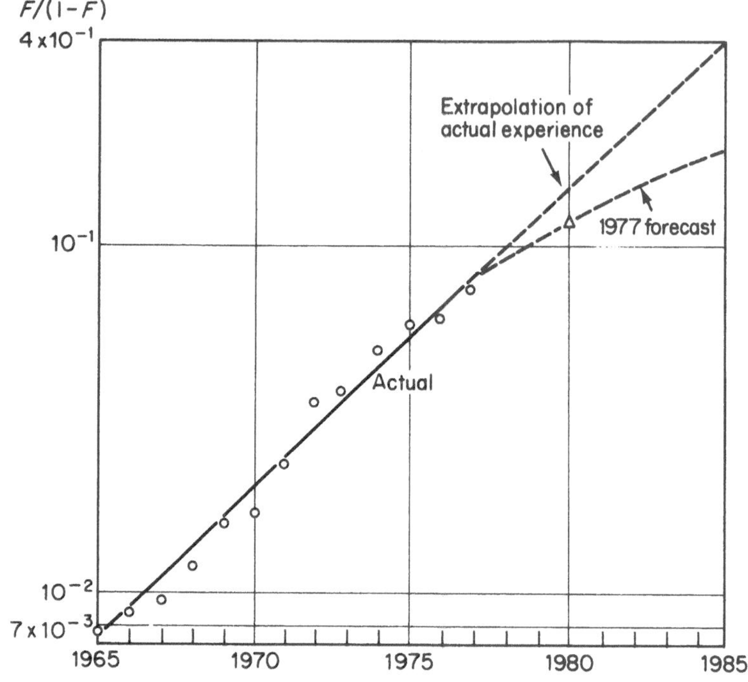

Figure 32: *Penetration of the electric power market by nuclear power (Fisher-Pry model).* F = *fraction of total electrical capacity comprised by nuclear. (Source: Ref. 6)*

fact had the opposite effect. Such an outcome was indeed foreseen by L.G. Brookes[7] who showed that the critical factor determining nuclear power growth is the rate of expansion of the electricity system as a whole. "It is no help to nuclear power – quite the reverse – if high oil prices simply make an already good case stronger whilst reducing the size of the market." But the problem of the 1980s and 1990s will not only be the twin economic difficulties of recession and inflation arising from higher oil prices: the more serious concern will be the physical shortage of oil.

It will then only be through a reduction in economic activity – no matter how this is brought about – that the demand for oil can be contained to match a stationary or even declining output. This indicates the alarming possibility that the world economy will find itself pushed into a downward spiral of recession by an energy shortage led by the shortage of oil. As energy usage is forced down to the level of the supplies that are available, the falling industrial output and rising unemployment will further restrict the demand for energy and hence the need for new electricity generating plant. Yet electricity from nuclear power, and increased coal supplies are the only means immediately available for increasing energy supply during the next 20 years. To escape from this "energy trap" it will be necessary, by deliberately planned government policies, to increase to the greatest possible extent the substitution of oil by electricity and to develop the wider use of nuclear power and coal.

There is no alternative. It is not possible for governments to buy their way out of trouble as they would try to overcome a recession by printing money. Energy cannot be counterfeited: it must be paid for in real, inflation-proof currency.

The cost of procuring new energy supplies and of saving energy will be high. For the European Community, Dr Brunner, the Energy Commissioner, has warned of the inevitable collapse threatening the world economy if western countries do not completely alter their energy policies so as to reduce spending on oil imports. To do this the European Community would have to invest over the next ten years some 50 billion DM per year to establish replacement energy. The energy investment would then go up from the present figure of 2–3% of GNP to about 12% of GNP. The possibility of achieving this, once a recession has taken hold and GNP is stationary or falling, must be considered remote. It is therefore essential to avoid further delays in developing the use of coal and nuclear power and for the substitution of oil by electricity. A programme of positive conservation, involving the replacement of existing equipment with new, more energy efficient equipment, motor vehicles, domestic appliances, industrial machines etc. while it is a necessary step would also be capital intensive and does not of itself provide an escape. Unless early action is taken the only remaining course will be one of negative conservation – making do without.

Replacement energy can only come from coal and nuclear power, and these will principally find their use through electricity. In addition to financing the expansion of these energy sources there must also be a programme of substituting electricity for oil. As Professor Dudley Jackson[8] has pointed out this will call for imaginative planning. The UK should use the oil revenues from the North Sea in an investment programme aimed at (1) building up capacity for electricity generation, from both coal and nuclear fuels; (2) a concurrently phased electrification of the transport system both between and within cities; and (3) carrying out the requisite research and development. As additional electricity supplies become available there could even be a temporary subsidy on electricity used as a substitute for oil. To achieve such a substitution the obvious response given the present long lead times is for a programme of nuclear power plant construction to be undertaken now, ahead of the short term requirements, so that electricity would be available by the 1990s.

The announcement on 18 December 1979 by the UK Secretary of State for Energy of a new nuclear programme, starting in 1982 to install up to 15,000 MW over a 10-year period is a step in the right direction. But even taking an optimistic 7 or 8 years to bring the new stations into operation this programme will only make a contribution to energy supplies over the period 1990–2000. By that time the short lived period of self-sufficiency from North Sea oil will have come to an end and the UK will be competing in a hard world market for scarce energy imports. What is required is a positive decision by governments to finance an early construction programme in anticipation of future demand. But there is little doubt that in some countries such proposals would run into determined opposition from those who are opposed to both nuclear power and a wider use of electricity, to the extent that many governments would shrink from embracing a controversial course of action, while their political opponents would espouse the anti-nuclear cause in seeking wider popular support. This situation is exacerbated in countries with strong provincial governments or regional separatist movements. A supra-national body such as the Commission of the European Communities or even a federal government as in Germany may propose an expanded nuclear policy, but the regional governments faced with approving and licensing nuclear sites, and closer to the sharp end of public opposition may be reluctant to follow the national policy, while a separatist movement, as with the Basques in Spain, can use an unpopular nuclear station "imposed" by an "autocratic" national government as a means of rallying local support. What then is to be done? Some would argue that it is better to postpone any confrontation with public opinion; that the time spent in trying to educate the public on nuclear issues could be regained by a readier acceptance at a later date, enabling site licensing and building times to be shortened. Others have proposed elaborate – and time consuming – public inquiry procedures so that the public can feel that they are able to take part in and exert a real influence on any decision before it is taken. Some governments have already had, or will have, referenda on the use of nuclear power – Austria decided against, Switzerland for, in each case by a narrow majority: Sweden and Denmark will hold referenda in 1980 and 1981.*

While delay is always an easy option, the risk and high cost to the economy in terms of loss of output, unemployment and social unrest, that would follow any severe or prolonged energy shortage would seem to be too high a price to pay for appeasing the opponents of nuclear power, if indeed this is possible. In many countries regulations on the use of energy – electricity for flood lighting and resistance heating; oil for domestic heating where the indoor temperature and heating period are controlled – are already introducing new areas of compulsion, with penalties for infringement. Rationing and its accompanying black market will undermine the social fabric. At the international level competition over limited or deliberately restricted supplies of oil could lead to the richer countries taking a greater share of supply at the expense of the poorer developing countries which can only increase international tension while in the end competition between the industrial countries themselves could lead to war.

Governments should then, wherever possible, take what steps are readily available to increase the rate of construction of nuclear stations; to promote the use of electricity and nuclear power; to develop supplies of coal; and to encourage the substitution of oil by these available forms of energy.

No country can afford to renounce the nuclear options. Alternative energy supplies may not be available.

With this conflict between the growing need for nuclear power and the opposition to building new nuclear stations it is difficult to predict the outcome. Clearly the longer the delay in restarting or accelerating programmes in countries such as Germany, Austria, Sweden, USA, Japan, the larger will be the energy deficit in the last decade of this century.

If plans are held back until, perhaps in two to four years time, recurring energy and electricity shortages drive home the lesson that nuclear power must be fully utilised it may be too late to do so: the grip of a deep recession could impede attempts to expand nuclear power production at a time when it is most needed. Unless prompt action is taken now the answer to the question "How much nuclear power can be expected to be built by 2000?" may well be – not enough.

* In the Swedish referendum on 23 March 1980 the two pro-nuclear alternatives calling for full exploitation of 12 nuclear reactors polled 58% of the votes. The Danish referendum now seems likely to be put off.

REFERENCES

1. Chauncey, Starr, European Nuclear Society Conference, Hamburg, May 1979.
2. World Energy to 2020, World Energy Conference.
3. *Energy Technology for the UK*, Department of Energy, December 1979.
4. Connolly, Hansen, Jaek and Beckurts, *World Nuclear Energy Paths*, The Rockefeller Foundation/The Royal Institute of International Affairs, 1979.
5. Dollezkal, Koryakin *et al.*, Paper No. IAEA-CN-36/334, Nuclear Energy Centres: Economic Environmental Problems, IAEA Salzburg Congress, May 1977.
6. P. Hogroian, "Projections of energy requirements and their implications", Uranium Institute Conference, July 1978.
7. L.G. Brookes, Nuclear Power Implications of OPEC Prices, *Energy Policy*, June 1975.
8. Dudley Jackson, "A World without Oil", *New Society*, 17 August 1978.

7
Nuclear power utilisation

The principal immediate use of nuclear power is in the generation of electricity where the infrastructure already exists for production, distribution and supply. It does not require any major administrative or organisational changes for the electric utility companies to add nuclear stations to their systems. The utilities are relatively few in number, in some cases national corporations – France, UK, Italy, Korea – with experienced technical management and staff, already involved with government regulatory bodies, and accustomed to working closely with suppliers and inspection authorities in maintaining a high standard of reliability of operation.

In addition to electricity generation the direct supply of low temperature process heat or steam, either from combined heat and power units from large central power stations or from small purpose built reactors will provide another use for nuclear power. These may take slightly longer to develop, with the need to build up heat distribution systems, where they do not already exist, and to establish an organisational and administrative structure to manage a number of small units being operated by a variety of users – industry, municipal etc.

In the medium term, after the turn of the century, high temperature heat from gas cooled HTR (High Temperature Reactors) or VHTR (Very High Temperature Reactors) will be used directly as heat input for metallurgical processes, coal gasification etc. In the longer term when oil supplies are nearing exhaustion nuclear power must be shown to be capable of providing energy in a form suitable for the complete replacement of oil. This could be through the production of hydrogen by thermo-chemical decomposition of water or by electrolysis.

Specialised application of nuclear power could come with nuclear ship propulsion and floating or barge mounted power stations.

ELECTRIC ECONOMY

In the face of the forthcoming oil shortage and higher price it is necessary to consider how the substitution of oil by electricity can be accelerated so that oil can be increasingly reserved for premium applications such as aviation fuel and as a raw material for petro-chemicals or even for synthetic protein.

A useful study of the potential for the substitution of oil by electricity in the US has been made by Westinghouse. The contribution of the different primary energy sources to the total energy use in the USA in 1972 is given in Figure 33.

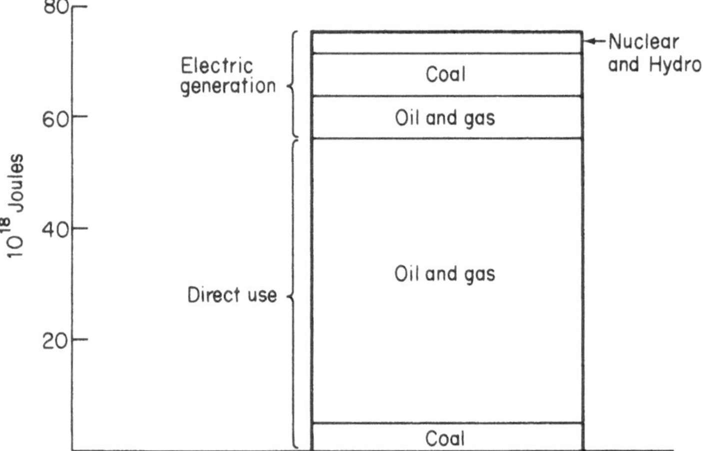

Figure 33: *US total energy usage, 1972*

The picture is of a fossil fuel economy with the major part being taken by the direct use of oil and natural gas. The US energy problem has then been defined as closing the gap between supply and demand curves for oil and gas.

A breakdown illustrating the pattern of the direct oil and gas usage in the US in 1972 (Figure 34) identifies transport as the main use of oil followed by space heating and process steam. It is in these areas that a substitution must be made by electricity to achieve a significant reduction in oil consumption.

The major problem to be attacked is the use of oil for transport. A study by the European Commission indicated that the transport

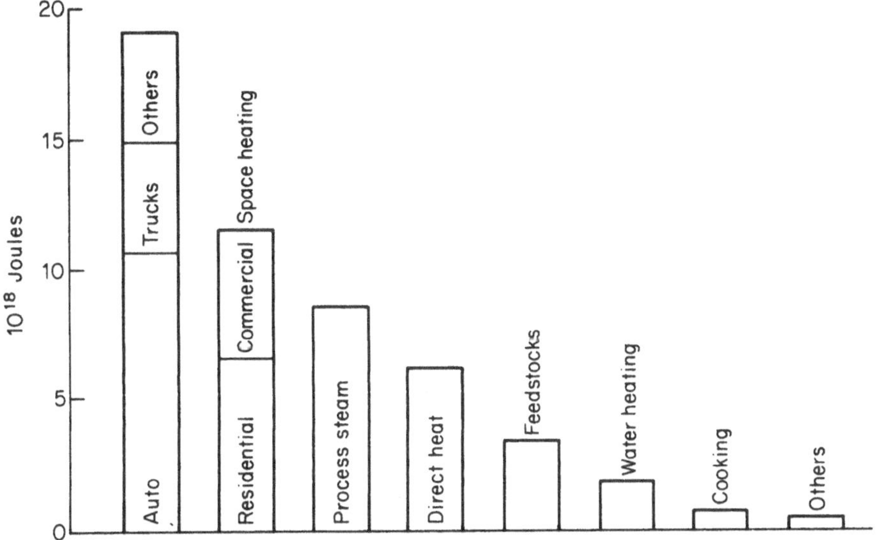

Figure 34: *Direct use of gas and oil in the US, 1972*

sector accounts for 28% of all the crude oil used in Europe, and of this private cars take nearly 70%. In the UK, the 1979 Energy Projections of the Department of Energy predict that, unless there is a breakthrough in transport technology oil demand for this sector could reach 35–40 million tons/year by the year 2000, which would account for over half of the energy demand for oil at that time. This figure includes the assumption that the fuel consumption of motor cars will improve from the present average figure of 30 miles per gallon to 40 m.p.g. In 1978 transport accounted for 23% of UK energy use.

It has been claimed that electricity could not only be a substitute for oil, but that it gives a more efficient usage of primary energy in a transport system. A comparison of an electric car with a petrol driven car shows that when the respective energy transformation processes are taken in account about 2½ times (14.1/5.3) more fuel is required by the petrol driven car; the 6 to 1 efficiency advantage of the electric drive system over the internal combustion engine outweighs the conversion losses of the electricity generation station (Figure 35).

The concept of the battery electric passenger vehicle is not new. The first practical electric carriage was built by Robert Davidson of Aberdeen in 1837. By the end of the 19th century fleets of electric buses were operating in Berlin, London, New York and Paris. In 1899 the world's land speed record of 65.82 m.p.h. was established by an electric car. In 1902 a firm, Accumulator

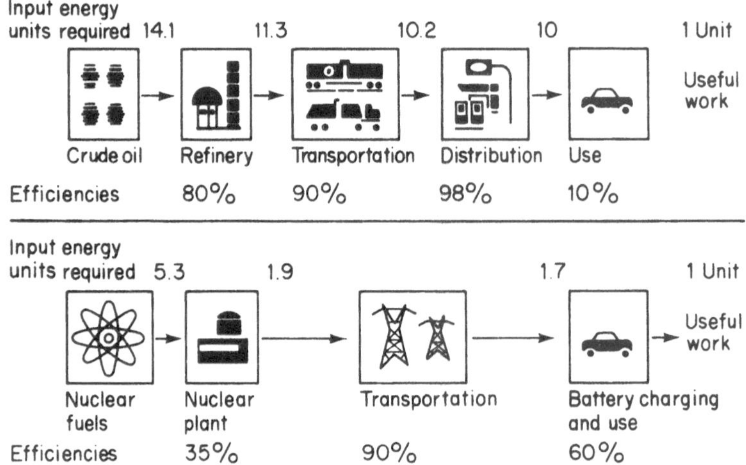

Figure 35: *Comparison of the electric and gasoline-powered car (Westinghouse)*

Industries of Woking in England, built a 12-seater coach with a maximum speed of 20 m.p.h. and a range of 80 miles per charge. Major advances are now being made in electric vehicle technology; an increasing number of electric vans are used for local delivery. In the UK there are over 40,000 battery electric (driver seated) road vehicles in operation, mainly for commercial deliveries. Electric buses are being developed for urban transport and attention is being directed towards electric cars mainly for use in cities. But to achieve any substantial substitution for oil will require a major effort on product and market development to be started without delay. In the UK the Electricity Council has taken a leading part in promoting the development of night-charged electric battery-driven motor vehicles. A fleet of 66 experimental vehicles – the Enfield 8000 – has been in operation for two to three years. These are two-seater saloon cars with a maximum speed of 40 m.p.h., 0–30 m.p.h. acceleration 12.5 seconds and a range on full battery charge of between 24 and 56 miles depending on driving conditions. By April 1978 the total fleet mileage (66 cars) was 193,266 miles, with one car being driven over 12,000 miles. The average energy consumption is in the range of 500–600 Wh/mile. This could be reduced with the use of light weight traction batteries. The battery is the key item. Costs of servicing are dominated by the battery replacement cost and battery depreciation is the major factor in vehicle running costs; it is 5–10 times the cost of electricity. One battery tested gave a total life of 3,832 miles, which corresponded to a cost of £0.069 per mile. With petrol in the UK now costing over £1 per gallon and likely to rise the electric car is almost competitive in straight economic terms. It is then necessary to develop batteries which will provide an improved range and performance and with a lower cost per mile over their life.' One aim is a battery delivering 30–40 kW of power at peak and 3–5 kW at steady load, with a storage capacity of 20 kWh. Charged overnight 10 million such batteries would consume 100 TWh per year – about one quarter of the UK electricity demand forecast for 1990. Such a large off peak load would contribute to improving the reliability and reducing the maintenance of nuclear plant which operate better at steady continuous running conditions. While the first electric cars will use advanced lead–acid batteries work is now being carried on in many countries on developing new lighter battery systems; these include alkaline nickel batteries which offer some weight advantage and a long cycle life; metal air batteries; alkali metal/sulphur batteries which operate a temperatures of 250–350°C; and a fluoride system. General Motors has recently claimed the successful development of a zinc–nickel oxide battery and is now considering building a small urban car or delivery van that could go 100 miles between battery charges, run at 50 m.p.h. and travel 30,000 miles before needing replacement. Such a vehicle might be commercially available by the mid-1980s.[2] Electric traction could also be used for city buses, freight and passenger trains. British Rail has announced its intention to extend railway electrification beyond the present figure of some 2,200 route miles – excluding the London Underground system – by up to a further 4,000 route miles. By the year 2000 about two-thirds of all train miles will then be electric.

In France the national railway company, SNCF, already has 6,000 miles of track electrified, and electrification is being continued at the rate of 100 miles/year with the aim of increasing the proportion of traffic hauled electrically to almost 90% by the year 2000.

An increasing reliance will be placed on electric trains for both freight and passenger traffic. In terms of energy use rail is four times more efficient than freight transport by road and seven times more efficient than aircraft.

As a substitute for oil in space heating, the use of electricity for domestic and commercial heating and air conditioning can be expanded, in particular by the increasing use of heat pumps. The heat pump can be regarded as a refrigerator in reverse. It takes free heat from an outside source, such as air, upgrades it and discharges it into a building at a higher temperature via a warm air distribution system. Even outside air at a temperature of 0°C still contains a great deal of heat when related to the minimum operating temperature of most heat pumps which is of the order of −10°C. Other sources of heat could be water or the soil. Small inexpensive heat pumps are being promoted for single purpose use – heating or cooling – without reversing cycles. Such pumps can show higher efficiencies. A Coefficient of Performance (COP) of up to 3 can be achieved, with 3 kW of useful heat energy being released for every kW of electricity consumed. Heat pumps are also being used in commercial premises as part of an integrated heating, cooling and ventilation system. It is claimed that up to 40% of the energy used for heating can be saved.

The use of oil for generating process steam can be reduced by electrode boilers. These would be particularly suitable for small scale intermittent use. Larger process steam requirements could be met increasingly from combined heat and power generation schemes.

/By converting to an electric energy economy the Westinghouse study estimated that by the year 2000, electricity generated from coal and nuclear power could provide some 75% of the total input for US energy requirement (Figure 36). This is of course

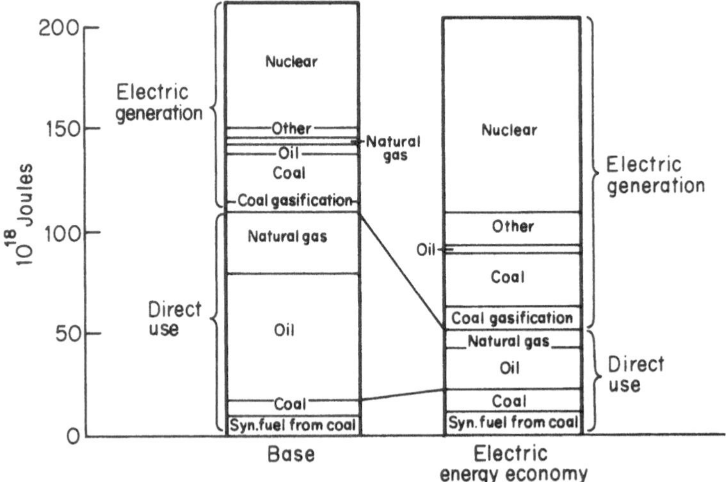

Figure 36: *US total energy economy, year 2000 (Westinghouse)*

an extreme figure obtained by pushing electricity applications to the full. It can be compared with the estimate by Chauncey Starr of the Electric Power Research Institute that by the year 2000 the electricity fraction of equivalent primary fuel input could grow from its present value to between 46–57%. But in the event of severe oil shortage this 75% figure could be taken as a goal for all the industrial countries. For the US the change would require 50% increase in electrical generating capacity which would go up from 2,000 GW to 3,000 GW by the year 2000 (Figure 37).

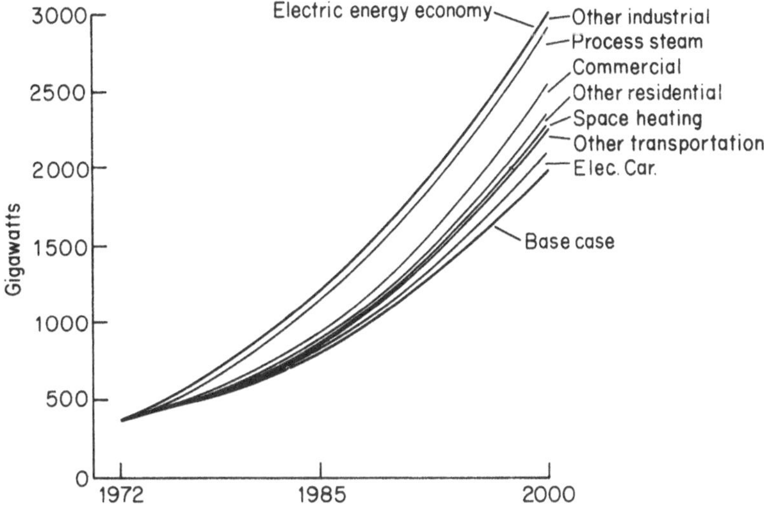

Figure 37: *US installed generating capacity*

Such a change could reduce the US consumption of oil to the level of the domestic production eliminating the need for oil imports.

REFERENCES

1. Data in this section from Philipp N. Ross, The Electric Energy Economy, *Westinghouse Nuclear Energy Digest*, No. 3, 1973.
2. *Financial Times*, 26 September 1979.

8
Nuclear process heat

The need to substitute oil and other fossil fuels by nuclear power has led to a growing interest in the direct use of nuclear heat. This could be available at temperatures ranging from hot water at 50°C to gas at temperatures up to 1,000°C. The water cooled reactors could provide heat at a temperature up to 300°C; the gas cooled High Temperature Reactors (HTR), using present technology, could extend the temperature up to 800°C, while the VHTR (Very High Temperature Reactor), for which considerable further development is required for materials of construction, reactor design and fuel, could provide hot gas at temperatures up to 1,000°C.

Figure 38 compares the temperature range required for a number of industrial processes with the temperature available from the different reactor systems.

LOW TEMPERATURE HEAT

In most countries the greater part of the heat demand is at temperatures below 200–300°C, within the range of the current design of light water reactor. For the UK and Canada 46% and 76% respectively of the industrial heat use is at temperatures below 300°C (these figures exclude energy used in transport, agriculture and homes). In Germany some 47% of the total energy is used as heat at temperatures below 200°C.

District Heating schemes have been in operation for many years. In Denmark for instance the first underground heat distribution networks were established as early as 1925. The combined production of heat and power using back pressure turbines is also a fully developed technology. It is then an obvious step to use a nuclear reactor as the energy source. This step has been facilitated by the development of improved and more economic heat distribution systems with better insulation and new piping materials. Plastic pipes, glass fibre or prestressed

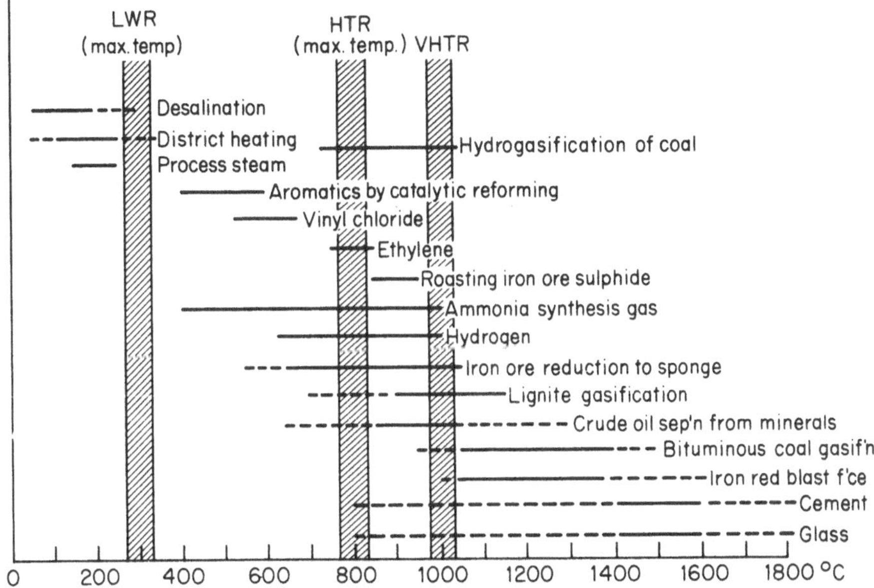

Figure 38: *Temperature ranges for process heat applications. (Source: Gulf General Atomic)*

concrete with a plastic lining are now being considered for the distribution of hot water at temperatures of up to 100°C and for distances up to 100 km. Heat from a large central plant transported for distances of between 40–100 km could be competitive with an on-the-spot fossil fired hot water boil (see Figure 39).

In Sweden, where most of the major cities already have heat distribution networks that are expected to undergo a considerable expansion to meet future demands, a study by a government committee considered the relative advantages of nuclear heat compared with fossil fired heating plant where the alternatives could be:

I: A single purpose oil fired hot water station which is likely to produce a mean long term SO_2 content in the atmosphere of 0.11 parts per hundred million (pphm).

II: A combined oil fired back pressure station with a hot water plant for peak loads which again would give an atmosphere SO_2 concentration of 0.11 (pphm).

III: A nuclear station backed up with a peak load hot water plant when the SO_2 content in the atmosphere (coming only from the peak load hot water plant) would be 0.025 (pphm).

In the absence of any district heating scheme local heating (L) would be used for say 80% of the inner city area when the SO_2

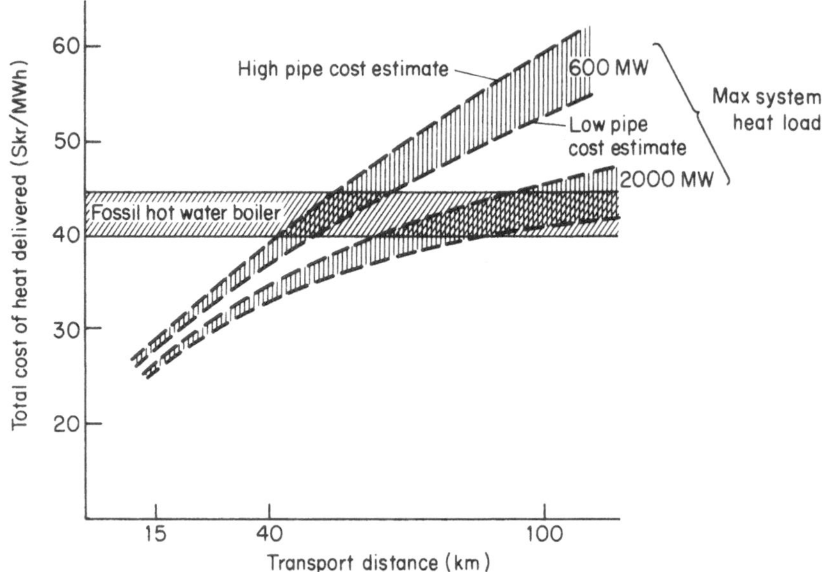

Figure 39: *Comparison of total cost of heat delivered for different maximum heat loads. Conventional technology. (Source: AB Atomenergi, Studsvik, Sweden)*

content would be 1.3 pphm.

The environmental damage was then estimated as in Table 54[1] which compares the effect of local heating schemes (L), and with the three alternative district heating schemes. The other sources of air pollution come from the remaining 20% of the city homes, industry, and the suburbs. These are listed under (R).

With the damage and health effects caused by sulphur oxides, particulates and nitrogen oxides, it was concluded that the fossil fired alternatives I and II would lead to a substantially higher mortality, illness and property damage than if a nuclear plant were installed, as in alternative III, but all three district heating schemes showed a marked improvement over local heating (L) and were much lower than the pollution from other sources (R). Detailed plans have since been worked out for a district heating scheme from the Barsebeck nuclear power plant for the cities of Lund, Malmo, Landskrona and Helsingborg in the south of Sweden. The nuclear plant would be a 3rd unit, Barsebeck-3 with a thermal rating of 3,000 MW. With a fully condensing turbine the electricity production would be 1,000 MW at an efficiency of 30%. By producing both heat and power, a heat output of 950 MW could be produced at the same time as generating 810 MW of electricity to give thermal efficiency of about 60%. The water would leave the power stations at 165°C with a cold return at 70°C.[2] (See Figure 40.)

Table 54: *Estimated environmental damages from different pollution sources in the model city*

Source		Excess mortality	Excess absence from work	Property damages
		cases/year	days/year	M SW.Kr/year
Heating, 80% of inner city area				
L	local heating	120	600,000	150
Alt I	district heating HVC	20	90,000	24
Alt II	district heating OKV + HVC	20	90,000	24
Alt III	district heating KKV + HVC	4	20,000	6
Other sources				
R	local heating 20% of inner city, heating of suburban houses, industries, distant sources	100	500,000	130

HVC – hot water centrals
OKV – oil fired heat and power units
KKV – nuclear heat and power station
Source: Ref. 1.

The most important environmental impact of this project would be to reduce the oil-based heat production of the area by about 500,000 tonnes of oil per year. This would reduce the amount of sulphur dioxide released to the atmosphere by 10,000 tonnes per year. The economic advantage of the scheme is shown in Figure 41.[2] The break even point for the first stage, Malmo and Lund only, compared with an extension of local oil-fired hot water stations is with heavy fuel oil at 280 Kr ton ($10/bbl), a price which was already exceeded by April 1975. For every increase of 10 Kr/ton in oil price above 280 Kr the competitiveness of the project increases by a total of 40 million Kr compared with an oil-fired hot water system.

In addition to combined heat and power from nuclear plant the production of heat only from purpose built nuclear heat reactors has been proposed. These would operate at low temperatures and pressures and could then be of a much simpler design and of a much smaller size than conventional power reactors. The low power density gives favourable operating conditions with a very low fission product release.

It is then possible to design a small reactor for safe operation in an urban or industrial site to provide heat at the point of use, thus

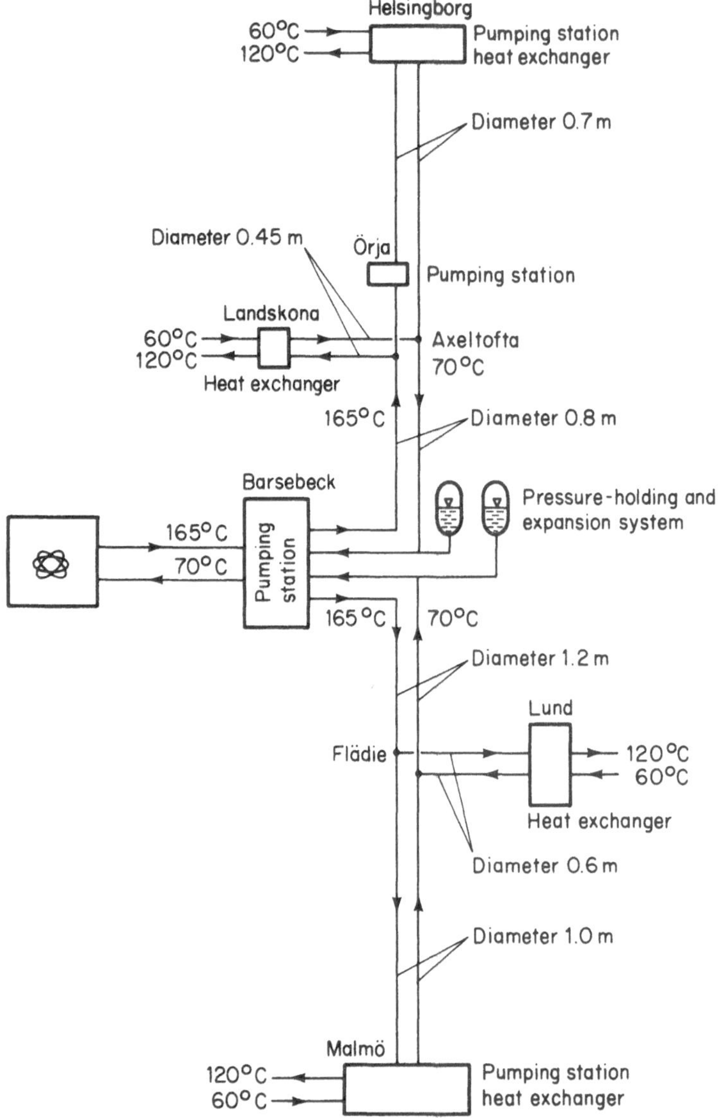

Figure 40: *Schematic diagram of transmission system. (Source: Ref. 2)*

avoiding the requirement for long heat distribution pipework from a more remote reactor site. Such a heat producing reactor could be built so that the shut down and core cooling are secured under all accident conditions without the need for any engineered safety

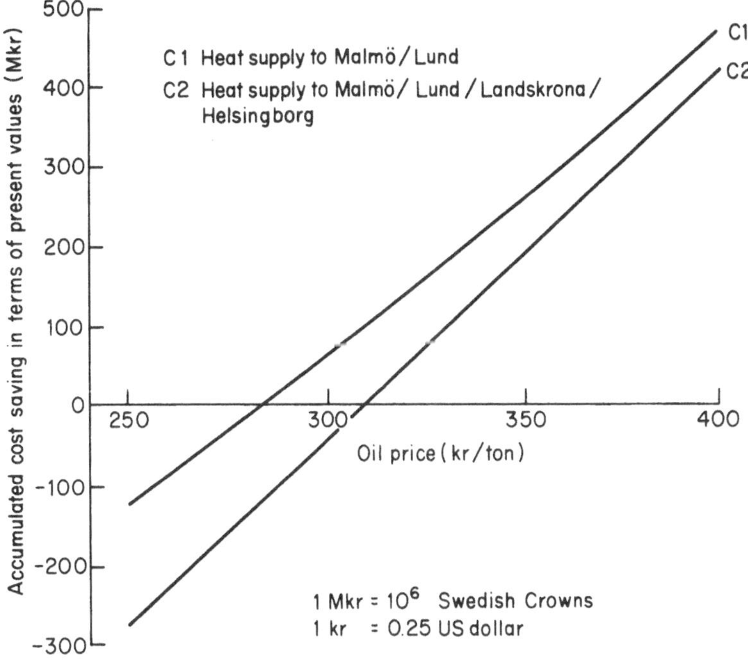

Figure 41: *Cost saving for transmission of heat from the Barsebeck power station compared with the extension of hot water stations. Variable oil price. (Source: Ref. 2)*

systems. These reactors could be sited above ground, recessed, or even underground, depending on local conditions and requirements. One such design by ASEA-Atom of Sweden (Figure 42) is for a 200 MW(th) reactor with an outlet temperature of 120°C at a pressure of 100 psi. The district heating supply temperature would be 100°C with a return at 70°C. The core would require 13 tonnes of uranium at an equilibrium enrichment of 2.7%. The core burn up at 27,000 MW days/tonne of uranium would give a theoretical life of nearly five years. Similar reactor designs have been proposed in other countries. The USSR in particular has plans to install several hundred heating reactors in major cities throughout Russia by 1990. In France it has been claimed[3] that by the year 2000 nuclear heat from small heat reactors and also from combined heat and power stations would contribute almost 2½% of the total energy requirements of the country -- 7 MTOE out of a total of 300 million. By the same date nuclear electricity would provide 100 MTOE; one third of the total energy.

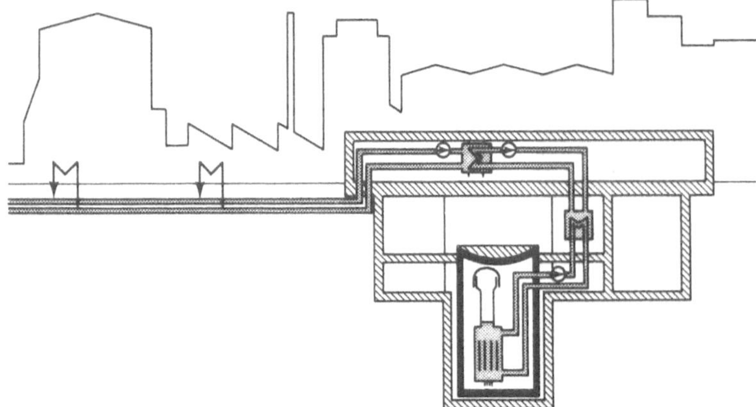

Figure 42: *Reactor providing heat where required. The plant can be adapted for aboveground, recessed or underground siting depending on local conditions. The compact layout features a reactor, primary cooling circuit and reactor auxiliary systems in the reactor building, and all conventional plant auxiliaries in the adjoining auxiliary building.*
(Source: Asea-Atom, Sweden)

HIGH TEMPERATURE HEAT

For many industrial processes heat is however required at temperatures above the range possible from an LWR (see Figure 38). These include metallic ore reduction and roasting, hydrogen production and the manufacture of synthetic natural gas from coal, some oil refinery and chemical plant processes. Suitable process temperatures could be obtained from a helium cooled high temperature reactor. Considerable development of the HTR has been carried out at the pilot plant stage, notably by the OECD Dragon project where a small reactor operated well with very low contamination of the primary circuit, and in Germany where a prototype pebble bed reactor is now being built. HTR development has however received a set back with the delays and difficulties encountered with the first commercial prototype at Fort St. Vrain in the US. The use of the HTR for process heat applications must now be considered as a longer term project which will probably follow HTR electricity stations. These are however unlikely to come into widespread use for at least 10–20 years.

With the present design of HTR steam could be generated at 510°C and 2500 psi with only small modifications to the reactor.

If the hot helium were to be used directly the upper limit is now about 700–800°C, but with further development temperatures up to 1,000°C could be obtained. There is then a growing interest in the heat which could be available from the HTR.

But there are many problems to be overcome before a commercial installation could be realised. One of the basic issues is the integration of large scale user plant with the nuclear power station which could have differing operational and maintenance characteristics. Safety, siting and plant reliability considerations are of particular importance, since the rigid linking of a high capital cost nuclear heat source to a large process plant will give an inflexible unit in the event of unexpected down times or output limitations from either of the component units. A critical item is the development of materials for construction of the heat exchangers to transfer the reactor heat to the process plant. It is a feature of nuclear plant that the facilities for maintenance are limited. But as high availabilities are required for industrial processes, the heat exchanger must be able to operate well within the limits of the properties of the materials used. The possibility of using high temperature ceramic materials such as silicon nitride is now being considered.

The first applications of high temperature nuclear heat could be in the manufacture of synthetic natural gas from coal. As fossil fuel reserves are depleting and coal becomes more costly, a high utilisation of coal carbon, and the conversion of coal to liquid and gaseous products will become increasingly important. A synthetic natural gas from coal also provides a more convenient use of the coal as a source of industrial and domestic heat. The combination of coal gasification using a nuclear heat source could then make a wide range of energy applications possible (Figure 43).

With coal as the prime substitute for oil it is necessary to conserve supplies to the greatest extent. The principal advantage of using nuclear heat in the steam gasification of coal, rather than generating the heat by burning additional coal is that the yield of methane would be increased and the amount of waste carbon dioxide emission would be reduced.

There are three possible processes for the manufacture of synthetic natural gas from coal:

(a) gasification with steam;
(b) gasification with hydrogen;
(c) gasification with hydrogen via a liquefaction process.

In all cases the process heat could be provided by burning part of

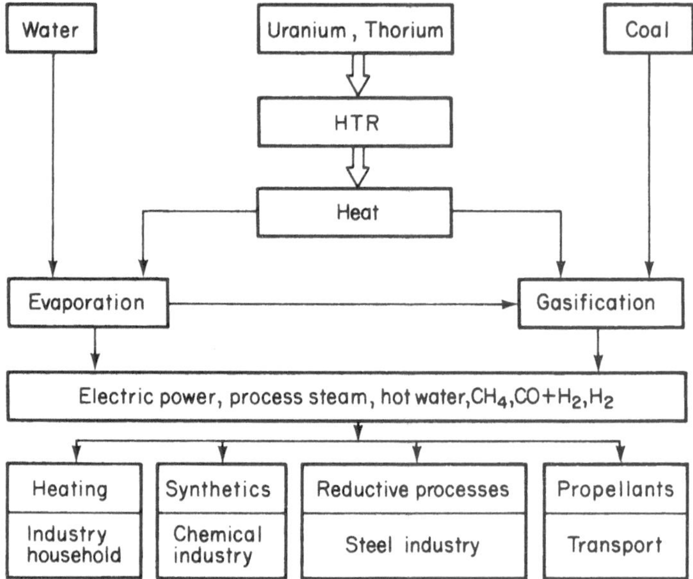

Figure 43: *Applications for nuclear process heat. (Source: Gulf General Atomic)*

the coal or by nuclear heat, which could also generate the steam in the first process and provide the energy for hydrogen production in the last two processes probably by the steam reforming of methane.

For the production of 1,000 m^3 of methane the requirements for the autothermal and nuclear heat processes would be:

Autothermal	Nuclear Heat Process
1.8 tons of coal	1.1 tons of coal
3.2 tons of water	2.9 tons of water
0.9 tons of oxygen	5.2 Gcal of nuclear heat

The main problem for the nuclear process is to obtain a sufficiently high temperature in the gasifier so as to obtain a high reaction rate and a high conversion. The heat is transferred from the reactor via a secondary helium circuit heat exchanger to protect the reactor coolant circuit from hydrogen diffusion and possible subsequent reaction with the graphite core, and at the same time to protect the process circuit from any radioactive products which might be present in the primary coolant circuit. A helium outlet temperature from the reactor of 900°C would give a gasifier temperature of 760°C, but the coal conversion rate could be

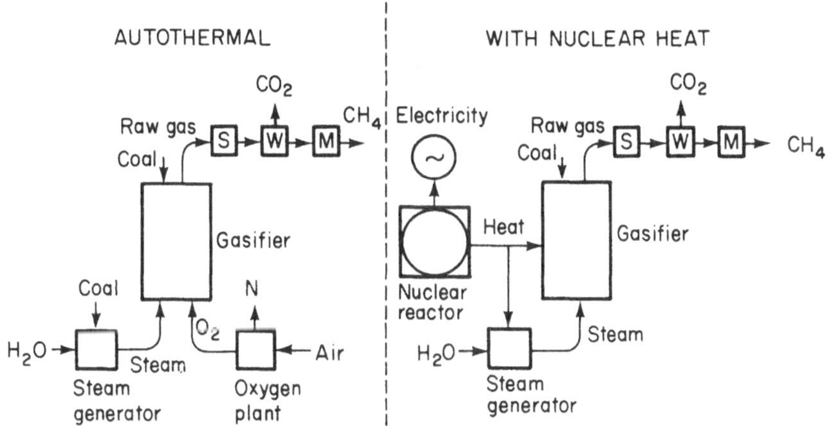

Figure 44: *Schemes of coal gasification processes. (S = CO-shift; W = gas scrubber; M = methanation)*

almost doubled if the gasifier temperature were put up to 800°C corresponding to a helium outlet temperature of 1,100°C.

Work has been done on this process in Germany. A semi-technical scale plant to gasify 200 kg/h of carbon with an electrically simulated helium circuit is being built and a demonstration plant coupled with a 750 MW(th) HTGR with a helium outlet temperature of 950°C has been proposed.

The two hydrogen gasification processes both depend on the production of the hydrogen by a methane steam reaction for which the energy is provided by nuclear heat at about 800°C.

$$CH_4 + H_2O \rightleftharpoons 3H_2 + CO:$$

Heat requirement 59.5 kcal/mol CH_4

The reaction could take place in a methane reformer within the pressure vessel of the reactor and heated directly by the primary helium. The hydrogen produced is then used to manufacture synthetic gas either in a single stage gasification process of the coal or in a liquefaction followed by a further hydrogenation.

Figures for a possible nuclear brown coal hydrogen gasification unit in Germany are:

Reactor Heat Output	3,000 MW
Brown coal throughput	12 million tons/year

Synthetic National Gas production	2,000 million m^3 /year
Residual coke	1.4 million ton/year
Electricity output capacity	760 MW
Nett electricity output	3,500 million kwh/year

Another process which consumes large amounts of energy is the production of iron and steel, where the units could be of sufficient size for the heat and electricity output of a reactor to be matched by the energy demand of the steel works. The process would require the production of reducing gas, by using nuclear heat in a coal gasification or methane steam reforming unit. The gas is then used to reduce the iron ore, a reaction which can take place at temperatures below 800–900°C. The iron could be further refined to steel in an electric ore furnace for which the energy would also be supplied by the nuclear unit.

In theory it would seem advantageous to take the hot reducing gas from the reforming unit and use this directly in the reduction plant, and this approach is being studied in Japan where a fully integrated design coupling the reactor and the iron reduction plant is being developed. For this a very high temperature reactor VHTR with a gas outlet temperature of 1,000°C is being studied by the Japan Atomic Energy Research Institute with plans to have a small 50 MW(th) reactor in operation in the 1980s. A supporting programme on iron ore reduction processes is being undertaken by the Japanese steel companies together with the Ministry of International Trade and Industry MITI. A small experimental integrated nuclear steel unit could be in operation by the mid-1980s. A process flow diagram for this system is shown in Figure 45.

With the requirement that the temperature of the reducing gas to the reduction furnace should be between 800–850°C, the reactor outlet helium temperature must be 1,000°C. For this studies on high temperature helium loops with a heat input of 3 MW(th) and a heat exchanger with a capacity of 1 MW(th) are planned. Small out of pile loops have already been operated and work is going ahead on the development of heat resistant super-alloys and of insulation materials which can retain their insulation properties in helium at a high temperature for a long time.

If this work is successful a semi-commercial pilot plant linked to a reactor of several hundred MW will be built leading finally to a commercial plant using a reactor of 3,000 MW heat output.

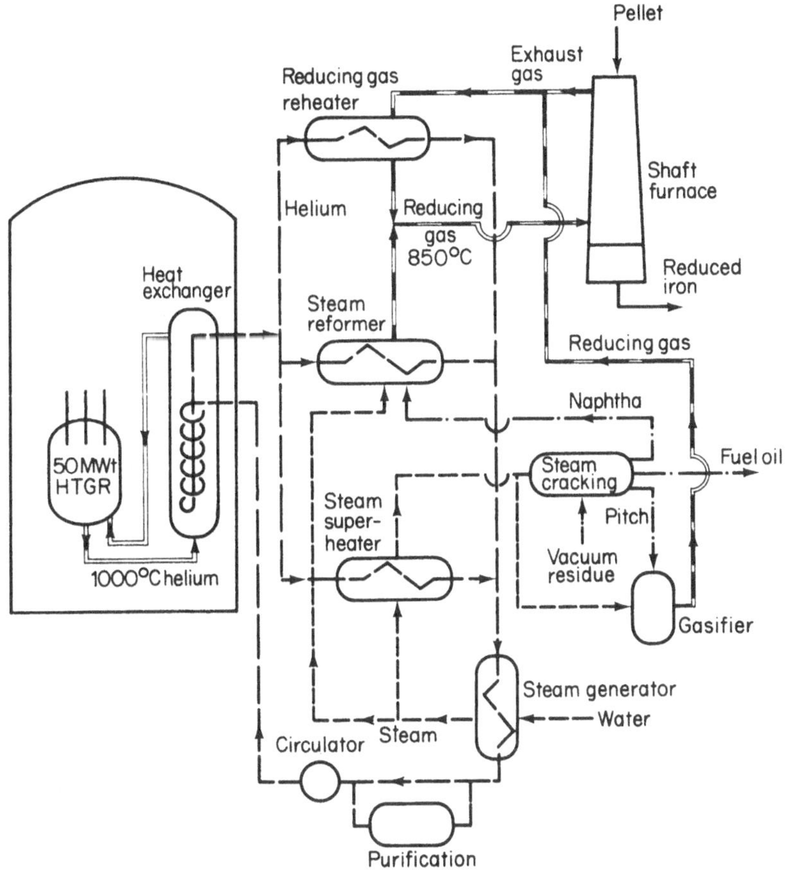

Figure 45: *Process flow diagram for nuclear steelmaking system.*
(Source: Japan Atomic Industrial Forum)

In Europe, the alternative of separating the nuclear plant and the steel works is proposed by the group of steel companies who are studying the process in the European Nuclear Steelmaking Club (ENSEC). This involves some loss of energy in that the reducing gas from the nuclear plant would have to be heated up to the process temperature at the iron works – probably by auto combustion. On the other hand the greater flexibility whereby both the gas and the electricity could be supplied to the steel works by a pipeline and an electric grid system has compensating advantages.

THE HYDROGEN ECONOMY

There is now a growing awareness of the potential for the use of hydrogen as a fuel, either in combination with carbon as synthetic oil, and methanol, or with nitrogen as ammonia, or used directly. The first step will probably come with the upgrading of coal by gasification or liquefaction processes which essentially take place by the addition of hydrogen to coal. Another end product could be methanol. These processes can be regarded as extending the fuel use of coal which can then be regarded almost as an energy carrier rather than a primary source of energy. Conventional coal liquefaction requires 2 units of carbon to produce 1 unit of methanol. High temperature nuclear process heat improves this ratio to 2 units of carbon to give 2 units of methanol, but with a hydrogenation process 2 units of carbon could give 4 units of methanol. This increases by a factor of 4 the conversion of carbon into a liquid fuel.

If the price of coal rises, or shortages occur, or if the CO_2 content of the atmosphere rises to unacceptably high levels with the danger of bringing about severe climatic changes, it will be necessary to turn to the direct use of hydrogen, alone, or combined with nitrogen as ammonia.

The use of hydrogen as an aircraft fuel is already being seriously considered. An international conference held in Stuttgart, September 1979, sponsored by Lockheed Aircraft of the US and the West German aerospace industry has proposed an experimental programme based on the use of four aircraft by an all-cargo airline to study the practical possibilities of making, storing and using large quantities of liquid hydrogen fuel in civil aviation. The aim is to promote hydrogen as an airliner fuel in the 1990s and beyond.[4]

Hydrogen has many advantages as a non-polluting fuel. The ignition energy of hydrogen is less than one-tenth that of natural gas so that it can be readily burned in low temperature catalytic burners. Under these conditions there would be no formation of nitrogen oxides. When burnt in pure oxygen the only combustion product is water. To produce hydrogen from the most widely available raw material, water, there are three methods: the steam/iron reaction, thermochemical water splitting cycles and electrolysis. The first two require high temperature heat, the third electricity. All these energy requirements could be met by nuclear power. Hydrogen is then not so much a new source of energy but rather an energy vector, a carrier which is used to transfer energy from a nuclear power reactor to the end use. In this way it offers a practically unlimited non-polluting source of energy and the hydrogen economy has been proposed as a solution to the long term energy requirements of the world.

Of the three possible hydrogen production methods, the steam/iron process is least attractive as it requires larger amounts of energy at higher temperatures. Electrolysis is already established as an industrial process and has the advantage that it could use off peak electricity and thus provide a means of storing energy from nuclear stations which are most efficiently operated at continuous full load.

The efficiency of existing electrolysis plants, based on the ratio of the amount of hydrogen combustion energy available to the electrical energy put in, is about 60–70%. Developments based on fuel cell technology suggest that improvements of up to more than 90% can be foreseen. Some high pressure prototype electrolysers have already reached 85% efficiency. The overall efficiency of an electrolyser process is however limited by the efficiency of generation of electricity which would be in the range of 35–45% to give an overall figure of about 30–40%.

With further development and larger scale manufacture, the capital costs of the electrolyser and the electrical conversion equipment, transformers, rectifiers etc. could be reduced by a factor of three or four from present levels. With off peak electricity for the electrolysis costed at marginal prices, electrolytic hydrogen could start to become competitive with oil at prices above $40/barrel.

In a search for more efficient hydrogen production methods, proposals were put forward by De Beni and Marchetti at the Euratom Research Centre at Ispra, Italy for a series of thermochemical reactions which could in theory give an overall production efficiency of more than 50%. The first of these using calcium bromide and mercury was:

$$CaBr_2 + 2H_2O \rightarrow Ca(OH)_2 + 2HBr \qquad \text{at } 730^\circ C$$
$$2HBr + Hg \rightarrow HgBr_2 + H_2 \qquad \text{at } 250^\circ C$$
$$HgBr_2 + Ca(OH)_2 \rightarrow CaBr_2 + H_2O + HgO \qquad \text{at } 100^\circ C$$
$$HgO \rightarrow Hg + \tfrac{1}{2}O_2 \qquad \text{at } 600^\circ C$$

The overall reaction if all the steps are combined is:

$$H_2O \rightarrow H_2 + \tfrac{1}{2}O_2$$

The reaction steps can be more readily seen in the representative series:

$$A + H_2O \rightarrow AO + H_2 \quad \text{at T1}$$
$$AO + B \rightarrow AB + \tfrac{1}{2}O_2 \quad \text{at T2}$$
$$AB \rightarrow A + B \quad \text{at T3}$$

In all potentially viable processes a change in the oxidation state of one or more chemical elements, either anions or cations, will be necessary. Possible anions include sulphur, the halogens and perhaps nitrogen and phosphorous. The cations of interest include nickel, vanadium and other multivalent metals or complexes.

A large number of alternatives are now being studied in many countries to find a series of reactions for which the upper temperature does not exceed 800°C, and the formation of corrosive intermediate products, requiring expensive materials of construction, is avoided. The kinetics, equilibria and degree of completeness of the reactions must also be established. A complex system of reaction vessels and heat exchangers must be devised to reduce the heat losses from the system. The physical nature of the reaction products has also to be considered, as the formation of a sticky solid product would involve serious handling and transfer problems for the separation and regeneration steps. These separation processes must not themselves require large amounts of energy which could adversely affect the efficiency of the system.

The cost of hydrogen from such thermolysis processes has been estimated at about \$5–7/GJ, which is rather less than electrolytic hydrogen using off peak power and an advanced conversion cell and corresponds to oil at \$35–45/barrel.

Other processes have also been suggested for hydrogen production. These include photochemical processes based on sunlight and using algae. Of some interest in the nuclear power context is a proposal to use ultraviolet radiation from the plasma of a fusion reactor for the direct photolysis of water vapour.

Distributing and marketing the hydrogen should not present great problems. The transmission of hydrogen in large diameter high pressure (up to 1,200 psi) pipes is well known in chemical plant installation. Pipelines carrying hydrogen over distances of up to 50 miles exist in Texas and South Africa, while a hydrogen pipeline network extending 230 miles has been in operation in the Ruhr since 1940. Pipeline transmission and distribution systems are widely used over long distances for natural gas. A hydrogen network would be similar; the costs have been estimated as some 20–120% more than for natural gas. Even at the higher figure the cost of transmitting energy in large quantities over very long distances by hydrogen pipeline may be less than that of transmitting the equivalent energy as electricity.

Hydrogen pipelines have been suggested as the most convenient way of transferring the power from a 20–30 GW nuclear complex perhaps sited at a distance from the load centre, or even offshore as a floating nuclear plant. The large volume of a hydrogen pipeline network which could also offer a means of energy storage provides another advantage over electricity generation where capacity must be available to meet instantaneous demand fluctuations.

If hydrogen can be produced from water as anticipated at a cost of \$5–8/GJ, it will be competitive with oil at about \$30–40/barrel. There would then be every incentive for making an immediate start on the large scale production of hydrogen and establishing a system of distribution, initially by using the electrolysis process and off-peak electricity. As experience grows, costs would be expected to fall so that eventually hydrogen could be produced at a competitive price at full electricity cost. It has always been considered that the first large scale use of hydrogen as an energy source would be in the preparation of synthetic liquid or gaseous fuels by the hydrogenation of coal. But with the rising price and expected shortages of oil the direct use of hydrogen will also be developed for early application as a replacement for oil in industrial heating processes and for space heating in domestic, commercial and industrial premises. The use of low temperature catalytic burners for space heating and the advantage of a non-polluting fuel in an urban environment should make hydrogen a preferred fuel for this purpose.

The advantage of a non-polluting fuel could also justify the early introduction of hydrogen for road transport. Only small modifications are required to burn hydrogen in an internal combustion engine. It has also been claimed that such engines would have a 50% greater efficiency compared with a petrol engine. For aircraft, hydrogen has the advantage of a 2.5 greater energy content per unit mass of liquid hydrogen, compared with conventional jet fuel, part of this advantage would however be lost in the increased weight of cryogenic storage tanks. The liquid hydrogen could however be used for cooling the aircraft surfaces at supersonic speed.

At present it is the chemical use of hydrogen, in particular in ammonia synthesis which represents the largest single use. With increasing world population, there will be a need for a greater food production which in turn will require increasing production of artificial fertiliser. One estimate puts the world hydrogen requirement for ammonia production at between 10,000 and 27,000 x 10^9 cu ft by the year 2000.

It has also been proposed that ammonia could also be used as a fuel as an alternative to hydrogen itself.[5] Since ammonia is already produced and distributed on a large scale as a nitrogen carrier it could equally well serve as a hydrogen carrier, thus complementing its agricultural role with the additional role as an efficient, low cost vehicle for energy storage and distribution. It would then be possible to use a single material for agricultural, chemical and fuel purposes with common production transportation and storage facilities. It is claimed that the costs of pipeline transport for ammonia, expressed in cents per GJ, would be appreciably less than for natural gas and even lower compared with hydrogen (Figure 46).

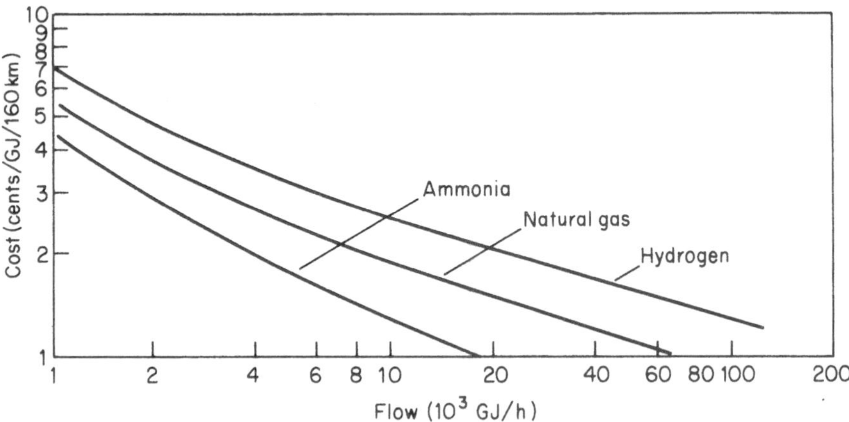

Figure 46: *Total pipeline transportation cost versus energy flow rate*

Ammonia can be readily burnt in conventional equipment, but would be most suitable for combustion in a gas turbine, where the temperature would be low enough to avoid the production of nitrogen oxides. As with hydrogen the main combustion product would be water, and this has led to the interesting suggestion that ammonia or hydrogen energy systems could also be used in arid regions to serve as a water distribution system to supplement other sources.

REFERENCES

1. Närförläggning av Kärnkraftwerk, SOU 1974:56, Industridepartment, Stockholm.
2. The Sydvarme Project, Sydkraft, fack S-200, Malmo 5, Sweden.
3. M. Chevrier, "Electricité de France", European Nuclear Society Conference, Hamburg, May 1979.
4. *Financial Times*, 18 September 1979.
5. L. Green, The Mitre Corporation, Virginia, USA.

9
The comparative hazards of energy production and use

There is a growing awareness of the risks and hazards associated with all sources of energy production and energy use; of the effect these have on man's environment, as well as directly on man himself.

In the past few years there have been a number of attempts to make a direct comparison between different energy sources. Much of this work is still at an early stage and there are considerable uncertainties over the effect of low concentrations of pollutants on man, but there is a growing consensus of informed opinion that nuclear power can be seen to be less polluting, less damaging to the natural environment, and safer for workers in the energy industries as well as the general public than most other forms of power.

ENERGY AND THE ENVIRONMENT

The environmental advantage of nuclear power compared with fossil fuels lies simply in the fact that much smaller quantities of materials are involved. One ton of uranium ore at 0.2% uranium content if used in a thermal reactor such as the LWR contains as much energy as 50 tons of coal; if used in a breeder reactor it would be equivalent to about 4,000 tons of coal. The effect of this quantity factor can be seen at all stages -- production, transport and end use for power generation. It bears directly upon the environmental nuisance as well as on the health and safety of workers and the public. These will be considered in turn, the comparison being based on the fuel required for a 1,000 MW electricity generating station operating at 75% load factor.

LAND USE

For coal a study in the United States[1] considers the land disruption of coal-mining. For stripmines the figure is put at between 0.06--0.32 hectares per 1,000 tons of coal. Underground mining, through subsidence and related effects can cause the collapse of buildings and prevent the use of the land for agriculture; it is estimated that roughly 0.04–0.08 hectares of land is so disrupted per 1,000 tons of coal mined. If it is assumed that half of the 2.3 million tons of coal per year for the 1,000 MW power plant are produced by stripmining and half by underground mining the total land disruption would be some 240 hectares per year.

Even higher figures are reported from Canada, where the total land disturbance for the mining of 21 million tons of clean coal in 1974 is estimated at about 16,000 hectares of which at most 20% can be reclaimed. On this basis the annual coal requirement of a 1,000 MW station would disrupt some 1,600 hectares of land, of which 1,280 would be permanently disturbed. The Canadian study[2] (Figure 47) also shows the very large quantities of waste; solid, liquid and discharged into the air which are associated with the coal mining operation. The figures are for the Canadian coal production in the year 1974, the quantities in millions of tonnes.

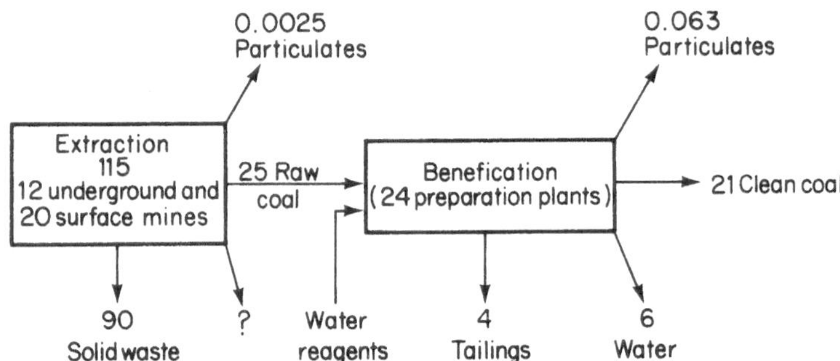

Figure 47: *Materials flow for coal mining and preparation in Canada. Values shown are nationwide annual totals in millions of tonnes, 1974. (Source: Ref. 2)*

The 94 million tonnes of solid waste is almost five times greater than the clean coal output, while the 6 million tonnes of water discharged from the beneficiation plant could also present an environmental problem as it is often contaminated with suspended and dissolved solids and of an acid or alkaline nature, although at some mines the water can be recycled. Atmospheric coal dust emissions at both the mining and beneficiation stages are also harmful, giving typical mean value figures of 79 $\mu g/m^3$ of air with peak daily values as high as 326 μg. The disposal of coal waste has lead to the creation of the numerous large slagheaps which so disfigure mining areas, particularly in Europe where the higher density of population makes them more objectionable. These slag heaps are not only unsightly but potentially dangerous. A coal waste tip which collapsed after heavy rain, at Aberfan in Wales in 1966, engulfed the village school killing 144 people, mostly children.

Mining and milling of uranium also produces wastes, but the quantities are very much smaller. The 130–170 tonnes of natural uranium feed per year for a 1,000 MW(e) power station could come from the mining of 65,000–85,000 tonnes of ore, with a concentration of 0.2% uranium. With richer ore the quantity mined would, of course, be smaller. This could be extracted in an open pit mine with a land use area of 6.4 hectares and the production of 2.5 million tons of overburden waste. Most of the land can be reclaimed after mining and piles of overburden can be graded, recontoured and revegetated. Milling and concentration of the ore gives rise to 86,000 tonnes of mill tailings. The mill tailings present a potential hazard, since they contain the thorium and radium present in the original ore. It is however possible to separate these elements, and this is done on a routine basis in some countries. An alternative is to cover the tailings pile with earth. This prevents their dispersion as wind borne dust and one metre cover of packed earth reduces the gamma ray activity by about 1,000 to negligible levels. The radium present will however produce measurable quantities of radon gas, although this is again reduced by earth cover. Revegetating the pile further decreases radon emissions by reducing the diffusion of the gas. An exclusion area of 1 km would reduce the hazard of those living in the vicinity to an insignificant level, since the radon concentration and external gamma radiation are at normal background levels except immediately over the deposits. Leaching of hazardous materials from the tailings pile by rain water is another potential hazard but this can be minimised by ditching around the pile to divert the surface run off water. In any case the danger of leaching by rain water is not necessarily a greater hazard than the leaching by ground water of the original ore body which could occur under the natural hydrological conditions before the ore was mined or even if it were not mined at all. A system of long term supervision or surveillance on a national level would be required to ensure that the mill waste material was not redistributed to the environment, although again for shallow ore bodies the hazard is not significantly greater than if the ore had not been mined. The land use for tailings piles would not exceed 0.8 hectares per year giving a total nuclear land use of less than 8 hectares per year for

Table 55: *Selected oil spills exceeding 2,000 tons in marine waters, OECD countries, 1967–1979*[a]

Year	*Polluter*	*Amount of oil released (tons)*	*Affected area*	*Type of accident*
1967	Torrey Canyon	117,000	UK/France	Went aground
1967	R.C. Stoner	20,000	North Pacific	Went aground
1968	Ocean Eagle	12,000	USA (Puerto Rico)	Went aground
1969	Santa Barbara Platform	6,000	USA (west coast)	Blew out
1970	Texaco Oklahoma	31,500	USA	Went aground
1970	Polycommandeur	16,000	Spain	Went aground
1970	Arrow	10,000	Canada (east coast)	Went aground
1970	Chevron Platform	10,000	USA (Mex. Gulf)	Caught fire
1970	Pacific Glory v. Allegro	6,300	United Kingdom	Collided
1970	Ocean Grandeur	2,500	Australia	Went aground
1971	Juliana	7,000	Japan	Went aground
1971	Oregon Standard	3,000	USA (west coast)	Collided
1973	Javvackta	16,000	Sweden	Went aground
1974	Mitzushima Refinery	8,000	Japan	Leaked
1974	Yuyo Maru	3,000	Japan	Collided
1974	Universe Leader	2,500	Ireland	Terminal Oper.
1974	Saglek	2,000	Canada (east coat)	Terminal Oper.
1975	Spartan Lady	20,000	North Atlantic	Sank
1975	Olympic Alliance v. HMS Achilles	2,100	United Kingdom	Collided
1975	Allied Chemical Barge	2,000	USA (east coast)	Sank
1976	Urquiola	100,000	Spain	Exploded
1976	Argo Merchant	25,000	USA (east coast)	Went aground
1976	Boelhen	10,000	France	Went aground
1976	Sealift Pacific	4,200	USA (Alaska)	Went aground
1976	Barge in Chesapeake Bay	2,700	USA (east coast)	Sank
1977	Grand Zenith	32,000	Canada (east coast)	Sank
1977	Ekofisk Platform	21,300	North Sea	Blew out
1977	Irenes Challenge	19,000	North Pacific	Sank
1978	Amoco Cadiz	230,000	France	Went aground
1978	Heleni V v. Roseline	4,000	United Kingdom	Collided
1979	Andros Patria	60,000	Spain	Caught fire
1979	Betelgeuse	35,000	Ireland	Terminal Oper.

[a] January and February only for 1979.

the 1,000 MW plant.* This is a factor of about 30 to 40 less than for coal in the USA and 200 times less than the figure for coal production in Canada.

Considerable advances have been made particularly in the UK by the National Coal Board in the reclamation of land after stripmining. The land so reclaimed can be reused for agriculture or, by landscaping, turned into a recreational area. Fresh water storage schemes, to be used as drinking water reservoirs or for

* The INFCE study estimates that for a 1,000 MW(e) nuclear plant at 75% load the land usage for a once through LWR system would be 12 hectares temporarily disturbed and 1.1 hectares permanently committed.

pumped storage, are another way of putting the reclaimed land to good use. Nevertheless some studies have concluded that land management could become a major problem if coal is to be extracted on the very large scale that will be required if it is to become a major substitute for oil.

Land use could also be a limitation to the large scale development of solar energy, at least in the populated, non-desert areas of the world. One estimate puts the land required for a 1,000 MW solar photovoltaic system as 72 km^2 of permanent occupancy. While for a solar thermal station the land area required is put at 35 km^2.[3] It is difficult to see where sufficient sites could be found in the UK or even the rest of Europe for any appreciable solar electric contribution to future energy supplies.

With oil, the main source of environmental pollution comes from oil spills. Table 55[4] lists 32 oil spills greater than 2,000 tons, in the seas surrounding the OECD countries over the period 1967 to early 1979.

With the development of offshore oil fields the hazards of marine pollution become greater, as well blow-outs are likely to become more frequent. Following Santa Barbara in 1969 and Ekofisk in 1977 the latest, 1979, blow-out of the Mexican IXTOC well, gushing oil at the rate of 20,000 bbls/day for a period of months, has become the world's biggest ever oil spill disaster. But while the large oil spills are more dramatic and claim greater attention, the smaller but continuous pollution from oil terminals and ports, as well as the oily water discharges from offshore platforms can be more significant in causing major changes to the environment in the affected areas.

REFERENCES

1. Wash-1223, *Comparative Risk–Cost–Benefit Study of Alternative Sources of Electrical Supply*, USAEC, December 1974.
2. Ripley, Redmann and Maxwell, Environmental Impact of Mining in Canada, Centre for Resource Studies, Queens University, Ontario, 1978.
3. A.M. Angelini, *Notiziario*, June 1979.
4. *The State of the Environment*, OECD, 1979, p. 63.

10
The hazards of fuel production

The main hazard in mining is the risk to the workers from accident and disease. Here again it is the much smaller quantities involved in the nuclear fuel cycle which swings the balance of safety heavily in favour of nuclear power. This is shown in a number of different assessments.

A comparison of coal and uranium mining accident rates based on actual US figures over the six-year period 1964–69 including both strip and underground mining, shows that they are remarkably similar when measured on the basis of fatal and non-fatal injuries per million man hours.[1] When allowance is made for the lower productivity of the uranium mines the accident rate per million tons of output is worse for uranium (see Table 56).

Table 56: *Average injury rates 1964–1969 for coal and uranium mines*

	per million man hours		
Uranium		*Coal*	
fatal	non-fatal	fatal	non-fatal
1.02	39.2	1.01	42.6
	per million tons of output		
Uranium		*Coal*	
fatal	non-fatal	fatal	non-fatal
1.39	53.4	0.45	19.0
	per 1,000 MW(e) power generated		

Source: Ref. 1.

But when calculated on the quantities of uranium ore and coal required to fuel a 1,000 MW(e) power station for one year the fatality rate is 0.1 for uranium compared with 1.0 for coal. A factor of 10 to 1 in favour of uranium.

The future trend for the uranium industry is however towards a more extensive use of open pit mining. Most of the large new discoveries in recent years are being developed by this method – Rossing in Nambia, the Ranger deposit in Australia and the Cluff Lake region of Saskatchewan. As the fatality rate of open pit mining for uranium is about one-tenth of that of underground mining the balance of safety can be expected to move even more in the direction of nuclear power.

Another set of figures, again based on US data, extends the comparison of the risk to workers to include oil drilling, production and refining (Table 57).[2] The period covered is 1965 to 1969 and the results expressed in disability days per 10^6 MWh electricity generated in 1969. For coal the calculation is on the assumption that 54.3% of the coal mined in the US in 1969 was used for electricity generation. The nuclear values were calculated on an estimated uranium ore requirement to fuel the actual nuclear generation in 1969. For oil 7.71% of the domestic production and 6.03% of the US refined petroleum was used in 1969 for electricity generation.

Table 57: *Comparative data on accidents occurring in various extraction processes from 1965 to 1969*

Process	*Accidents per year*		*Injuries per 10^6 man-hours*	*Disability days per 10^6 man-hours*	*Disability days per 10^6 MWh, 1969*
	Fatal	*Non-fatal*			
Coal mining	246	10,251	43.5	8,441	1,545
Uranium mining	8	272	39.8	8,702	157
Uranium milling	0.2	59	17.0	1,091	
Oil drilling and production	1,104[a]		10.2	1,176	135
Oil refining	1,060[a]		5.5	793	

[a] Includes both fatal and non-fatal accidents.
Source: Ref. 2.

These figures again show the 10 to 1 ratio in favour of nuclear power compared with coal. The risks of the oil industry which come out slightly lower than for uranium can however be expected to rise considerably with the expansion of offshore exploration and production. Accidents occur most frequently during the exploration and construction stage, placing the drilling platform on the sea bed, fitting out, equipping and laying pipelines. Crane operators and divers have been particularly at risk. Figures for the North Sea oil fields show that in 1976, with a work force of about 9,200 employed on offshore installations there were twenty fatalities. This gives an appallingly high death rate of almost 1 in 500, about ten times the risk of fatal accidents in British coal mines in the same year.

In addition to death or injury caused by accidents the longer term effects of industrial disease must also be taken into account.

In uranium mining the specific occupational risk is of lung cancer caused by the inhalation of radioactive radon gas. Radon is formed as a daughter product from the radioactive element radium and if the ventilation of the mine is inadequate the radon is inhaled and the decay products deposited in the lung can give a sufficient radiation dose to cause lung cancer. A clear relation between the frequency of such tumours in excess of the normal expectation, and the average radon concentration of the mines has been shown, but significantly the frequency of tumours was also strongly linked with the smoking habits of the miners concerned. In many uranium mines smoking underground is now prohibited. The figures from this pioneering survey, published in 1969 by F.E. Lundin, over the period 1950–1967 indicated an excess death rate from lung cancer in uranium miners of 15 per 10,000 man years worked. There are however good grounds for believing that this figure is higher than would now be expected under present conditions. The tumours recorded would only have developed many years after the exposure which caused them and this occurred in the very early days of the uranium mining industry. Once the danger was known the mining conditions and the ventilation rates were progressively improved. One estimate by Sir Edwin Pochin suggests that the average rate at the recommended levels now being achieved in most mines would probably be at about 1.5 excess deaths per year for 10,000 miners after an average of 20 years exposure at 4 "working level months (WLM)*" of radon concentrations per year – the US maximum permissible exposure level in mining. Other estimates of lung cancer mortality by a World Health Organisation working group put the figure for miners who have worked for 30 years in uranium mines as between 2.1 to 5.5 excess deaths per year up to the

* A Working Level is the concentration of radon and the decay products in equilibrium with it – the radon daughters – in one litre of air that would release 1.3 x 10^5 MeV of alpha particles during their radioactive decay. A working level month is the product of working levels in the inhaled air multiplied by the number of working months of exposure when one working month equals 170 working hours. The maximum permissible exposure for workers is 4 WLM or an average of 0.33 Working Levels over the year. For the general population in the home or elsewhere the limit is 0.02 Working Levels. The average exposure of workers at the El Dorado underground uranium mine in Canada is less than 1 WLM per year.

age of 80, per 10,000 miners, at the US limit of 4 WLM per year. For mining the uranium ore to fuel a 1,000 MW(e) reactor for one year the lung cancer risk can then be put at less than 0.01 cases per year, more probably between 0.001 and 0.005. Radon is a more severe hazard in the confined workings of an underground mine, and would not present the same problem with open pit mining.

The Board of Inquiry appointed by the Saskatchewan Government to review the Cluff Lake uranium mining and milling project,[3] which would be worked as an open-pit operation, estimated "as an extreme and unlikely upper limit" that the number of fatalities among the workers could be 450 per year per million workers. Of this 150 were deaths by industrial accidents. This figure of 450 per year per million can be compared with the fatality rate in other Canadian industrial operations (Table 58).

Table 58: *Accidental fatalities per million workers per year*[a]

Saskatchewan mines (uranium, coal, potash, hard rock)	680
Ontario manufacturing	66
Ontario construction	296
Ontario open-pit mines	320
Ontario underground mines	892
Ontario logging	1,122
Ontario logging, sawmilling and veneer milling	692
Fire fighters	10,000
Steel workers	600
Railroad workers	4,000

[a]For comparison, in the total population of Saskatchewan, the accidental fatality rate from all types of accidents is around 610 per million people per year, with approximately 300 of these being due to traffic accidents. The figure for Saskatchewan mines was obtained from the Saskatchewan Department of Labour, the Ontario figures were obtained from the Ham Report (1976).[3]

Underground coal mining can also lead to lung disease. The coal miners "black lung" is a form of pneumoconiosis due to the inhalation of coal dust that affects many older miners. Here again as in uranium mining the incidence of the disease can be decreased by improved ventilation and mining conditions. One estimate suggests a range of up to 3.5 premature deaths per year from disease associated with the coal production for a 1,000 MW(e) plant.[4]

A report by the British Health and Safety Commission puts a much higher figure, in the region of ten deaths per year from pneumoconiosis associated with the generation of 1,000 MW year of electricity from coal. But this reflects mining conditions over the past three or four decades and is expected to fall to one or two with the improved conditions in the mines.

These figures suggest that for disease, there is a factor of at least 30 in favour of uranium compared to coal.

In addition to occupational accidents the transport of fuel also involves the general public in traffic accidents. The figures are surprisingly high. But here again the quantity factor plays a large part in identifying uranium as the preferred fuel from a safety point of view.

For uranium the milling operation is usually carried out in close proximity to the mine with one central milling plant serving several mine workings. The low concentration of uranium in the ore, about 4 lb/ton, is one reason why milling is carried out near the mine. The further transport of the refined product as yellow-cake, U_3O_8, or at a later stage as enriched fuel or fabricated fuel elements is a matter of moving a few hundreds or eventually tens of tons of material per year.

For coal on the other hand transport to the power station can be a considerable hazard to the public as well as major cost item in the price of delivered coal.

In the United States 52% of coal delivered to the power stations arrives by rail, 34% by water and the remainder by road transport, conveyor belt or other means. The average distance between mine and power plant is 300 miles. The cost of transportation can amount to some 40% of the fuel cost to the utility. Each year in the US about 2,300 people are killed in railway accidents and since 20% of the total tonnage carried by train is coal (about one half of which is for power station use) approximately 230 deaths per year may be attributed to the transport of coal to power stations. Only 10% of these fatalities are railway employees. The major cause of accidents is by collisions with motor vehicles at level crossings. Since the electricity capacity represents the equivalent of 180 stations of 1,000 MW each, 1.3 deaths are associated with the transport of coal per 1,000 MW/year.[5]

Another estimate of the injuries attributable to the transport of fuel for a 1,000 MW coal and nuclear plant have been estimated for the public alone and are shown in Table 59.

Transport is also a source of danger and pollution in oil and gas industries. Collisions at sea involving super-tankers have led to wide scale pollution of holiday beaches and damage to fishing, particularly in the crowded sea lanes of the approaches to Europe. Loss of life has occurred not only in these collisions but also when discharging cargoes and in road accidents involving natural gas container vehicles. Two recent events were the explosion of the oil tanker Betelgeuse at the Irish terminal in Bantry Bay in 1979 when 50 people were killed, and the disaster at the Spanish holiday camp in 1978 when 150 people were killed and hundreds more injured when a gas tank lorry exploded.

Table 59: *Injuries due to transport of fuel for a 1,000 MW plant*

Public	*Coal*	*Nuclear*
Fatalities	0.55	0.009
Non-fatal injuries	1.20	0.080
Man days lost	3,500.00	60.000

Source: Ref. 1.

To complete the picture, figures should be given of the accidents associated with power plant operation and maintenance as well as for the stages of the nuclear fuel cycle. In the uranium fuel cycle the quantities of material are reduced even further. After conversion into uranium hexafluoride and passing through the enrichment process, the amount of enriched uranium to be fabricated into fuel, assuming an equilibrium annual reload cycle in which one-third of the reactor fuel is replaced each year, is only about 30–35 tons. All the fuel fabrication and reprocessing operations are either carried out under clean room conditions or remotely so that the risk to operating staff is low. Pochin puts the death rate from accidental injury for fuel fabrication reactor operation and reprocessing at a total of the order of 0.04 per year associated with the fuel cycle of a 1,000 MW reactor.

With this data it is now possible to put forward estimates for the total occupational risk of electricity generation. Three sets of estimates are given here, one from the US, one from the UK, and the third from an international World Health Organisation study group. The results are in striking agreement.

The US figures[6] present the expected injuries and deaths for 1980 fuel cycle for coal, oil and a LWR grouped under the three comparable headings of Fuel Production, Power Plant Operation and Maintenance and Transport. For both coal and LWR fuel cycles mining contributes more than 90% of the expected injury and death. With the big difference between open cast and underground mining it was assumed that each contributed half of the total production of both coal and uranium (see Table 60).

The UK figures (Table 61) were given in a 1978 report by the Health and Safety Commission, *The Hazards of Conventional Sources of Energy*.[7]

Table 60: *Expected numbers of annual injuries and deaths due to occupational accidents per 1,000 MW(e) plant for 1980 fuel cycles*

Cycle	*Fuel production*[a]	*Power plant operation and maintenance*	*Transportation*	*Total*
A. Injury (non-fatal)				
Coal	41.00	1.500	5.200	48.00
Oil	10.00	1.500	1.100	13.00
PWR	5.20	1.300	0.042	6.50
B. Death				
Coal	0.98	0.037	0.055	1.10
Oil	0.11	0.037	0.030	0.18
PWR	0.10	0.011	0.0017	0.11

[a]For coal, this category includes mining and preparation. For oil, it includes production and refining. For PWR, it includes all steps from mining through fuel element fabrication and reprocessing.
Source: Ref. 6.

This report comments that occupational mortality due to pneumoconiosis in coal miners and lung cancer in uranium miners are difficult to quantify and that the figures now experienced reflect the working conditions extending back over several decades. The figure for currently certified deaths from pneumoconiosis associated with the use of coal in electricity generation is in the region of 10 per 1,000 MW year but current mining conditions and the regular medical surveillance of miners are such that the figure from present dust exposures should be much lower, perhaps only one or two. These figures are however not included in the comparison of deaths due to accidents in Table 61. On the other hand the deaths from lung cancer in uranium miners are included: this figure is however small, not more than 0.01 per GW year.

A third set of figures is given by a Working Group of the World Health Organisation in their report on the Health Implications of Power Production. Mainly using data from a survey of statistical information on accidents involving death and injury for workers in the nuclear, coal and oil production industries, including mining operations by Comar and Sagan[8] and adding their own estimate for the nuclear industry the Working Group put forward Table 62.

Table 61: *Estimated number of deaths due to accidents per GWy[a] of electrical energy sent out[b]*

Primary energy source	*Operation*		*Deaths/GWy sent out Deaths caused by accidents*
Coal[c]	Extraction		1.40[d]
	Transport		0.20[e]
	Generation		0.20[f]
Total			1.8
Oil and gas	Extraction		0.30[g]
	Transport	Insignificant[h]	
	Generation	None reported[f]	
Total			0.3
Nuclear	Extraction (USA)		0.10[i]
	Transport	Insignificant	
	Generation and reprocessing		0.15[j]
Total			0.25

[a]Gigawatt years (giga = 10^9).
[b]Based on average electricity supplied from stations for years 1972–74 reproduced in Table 71, *Digest of United Kingdom Energy Statistics 1975*, which shows relative outputs for nuclear, oil-fired, and other steam raising plants (of which the vast majority are coal-fired).
[c]Based on underground mining figures only. Opencast fatalities are not included. (They would have an insignificant effect on the final figures.)
[d]Based on an assumed figure of twenty-six deaths in mining coal for power stations.
[e]Based on estimated number of deaths attributable to the movement of coal by rail not including accidents to the public. Figures for deaths due to the movement of coal by road considered insignificant.
[f]Based on number of deaths to CEGB employees 1970–77. Information supplied by CEGB.
[g]Based on figures for fatal injuries in exploration and production in the United States published in the 46th Annual Review of Fatal Injuries report to the American Petroleum Institute. It is appreciated that the bulk of the fuel oil used in UK power stations is from the Middle East where accident performance may differ sharply from that of the USA. Figures for the Middle East are not available and we have therefore used the American figures as the only available indicator. Making allowances for the lower fuel oil consumption of the UK we have arrived at a figure of approximately two deaths per annum in overseas oil fields attributable to the production of oil for UK. To these we have added the figures for deaths in the North Sea Oil extraction industry published by the Department of Energy. This gives an approximate figure of twelve deaths per annum.
[h]We have been unable to obtain figures for deaths due to the shipping of oil from overseas oil producing countries.
[i]Based on figures from United States uranium mining reproduced in USAEC Wash 1224 Comparative Risk – Cost – Benefit study of alternative sources of electrical energy. It is appreciated that the USA is not the primary supplier of uranium for UK power stations. Again we have used these figures as they are the only ones readily available.
[j]Based on information supplied to CEGB (see g above) and by British Nuclear Fuels Limited for accidental deaths of employees 1970–77. None of these deaths was due to radiation effects.
Source: Ref. 6.

Table 62: *Occupational accidental deaths from 1 year operation of a 1,000 MW electrical power plant and associated fuel cycle services*[a]

Fuel	*Occupational accidental deaths*[b]
Coal	0.54 – 5.00
Oil	0.14 – 1.30
Nuclear	0.01 – 0.86
Nuclear WHO estimate	about 0.3

[a]Data on accidents during construction of installations are not included.
[b]The ranges are the lowest and highest estimates for four studies cited by Comar and Sagan.
Source: Ref. 9.

REFERENCES

1. Wash-1224, *Comparative Risk–Cost–benefit Study of Alternative Sources of Electrical Energy*, USAEC, 1974, pp.3–69.
2. L.B. Lave and L.C. Freeburg, "Health Effects of Electricity Generation from Coal, Oil and Nuclear Fuel", *Nuclear Safety*, **14** (No. 5), 409 (September/October 1973).
3. Cluff Lake Board of Inquiry, Department of the Environment, 1855 Victoria Avenue, Regina, Saskatchewan.
4. L.D. Hamilton, *The Health and Environmental Effects of Electricity Generation — A preliminary report*, Brookhaven National Laboratory, 1974.
5. L.A. Sagan, Health costs associated with the mining, transport and combustion of coal in the steam-electric industry, *Nature*, **250**, 12 July 1974.
6. K.A. Hub and R.A. Schlenker, "Heath Effects of Alternative Means of Electrical Generation", IAEA-SM-184/18.
7. Health and Safety Commission, *The Hazards of Conventional Sources of Energy*, HMSO, London, 1978.
8. Comar and Sagan, *Annual Review of Energy*, 581 (January 1976).
9. *Health Implications of Nuclear Power Production*, WHO, Copenhagen, 1977.

11
The hazards of power generation

As well as the risks to workers in the fuel and power industries the hazard to the general public arising from different means of electricity generation must also be considered. The main source is in the discharge of wastes from the power station. Here again the smaller quantities involved weigh heavily in favour of nuclear power. There is in addition a major difference of attitude in that the policy of the nuclear industry is that waste should as far as possible be concentrated and contained, whereas the policy with regard to fossil fuel plants is that the gaseous discharges should be diluted and dispersed.

The quantities of waste products that arise when fossil fuels are burnt are large. Table 63 gives indicative figures based on the annual fuel consumption of a 1,000 MWe power plant at 75% load factor producing 6.6 billion kWh/year.[1]

Table 63: *Quantities of waste products produced from fossil fuels (tons)*

	Coal	*Oil*
Annual consumption of fuel	2.3 million	9.2 million barrels
Annual production of waste		
Bottom Ash	50,000	
Fly Ash retained	248,000	
Sulphur retained	46,000	
Annual discharges to atmosphere		
Carbon dioxide	6,000,000	4,500,000
Nitrogen oxides	27,000	22,000
Sulphur dioxide	24,000	21,000
Fly ash	1,000	150
Carbon monoxide	1,000	7,500
Mercury	5	
Arsenic	5	
Nickel	5	
Beryllium	0.4	
Lead	0.2	
Cadmium	0.001	

Source; Ref. 1.

This is but one set of figures. There will be quite large variations for coal and oil of different type and origin.

The major concern from the environmental and public health point of view is in the discharge of gaseous material. Although the policy of dilution and dispersal by discharging these wastes through tall chimney stacks lessens the harm in the immediate vicinity of the plant it is now recognised that this merely transfers the problem elsewhere. One of the major pollutants is sulphur dioxide, where the amount discharged into the atmosphere by all the OECD countries is put at 46.3 million tonnes. By 1985 this is expected in the "most probable forecast" to rise to 57 million tonnes. There are now active policies to try to prevent an increase in SO_2 emissions by preferential use of low sulphur oil, by coal washing and by flue gas desulphurisation. The potential for reduction is however limited and costly. For OECD Europe a reduction in the amount of sulphur dioxide discharged from 25 million tonnes/year expected in 1985 down to 9–10 million tonnes is estimated to cost $5 billion/year in operating costs alone, about 0.3% of OECD Europe GDP or $15 per capita per year. Even a smaller reduction from 25 down to 19 million tonnes would cost $1.8 billion.

While considerable emphasis has been placed on the SO_2 level in setting air quality standards in a number of countries, more recent work suggests that it is the oxidation products of sulphur dioxide – sulphuric acid and particulate sulphates – possibly acting synergistically with nitrates, and particulate matter which are the main cause of harm. Greater attention is also now being directed towards the discharges and health effects of the nitrogen oxides, hydrocarbon and organic chemicals and metals (Table 64).

Both coal and oil contain a number of harmful metals. Although the concentrations are low, the figures in Table 65 are in parts per million,[3] the annual consumption of over 2 million tons of coal for a 1,000 MW(e) power station brings the actual quantities released to potentially damaging levels.

Some of these, such as arsenic, chromium and nickel are carcinogenic – and there is a growing belief that most carcinogens can also cause genetic damage. Others, cadmium, copper, mercury and lead are toxic in varying degrees. Whilst some of these elements will be retained by the dust separation equipment, either electrostatic or cloth filters, others will to a lesser or greater extent be discharged into the atmosphere. Mercury for instance is almost totally discharged. Uranium and thorium are also present in many coals.

It should of course be recognised that fossil fired power stations

Table 64: *Annual releases of chemical pollutants from 1,000 MW(e) fossil-fuelled power stations*

Pollutants	Total annual emissions (per 1,000 t)	
	Coal[a]	Oil[d]
Aldehydes	0.052	0.1200
Carbon monoxide	0.520	0.0084
Hydrocarbons	0.210	0.6700
Nitrogen oxides	21.000	22.0000
Sulphur oxides	139.000 [b]	53.0000
Particulate matter	4.500 [c]	0.7000

[a]Represents the burning of about 2.1 10^6 t/year of semi-bituminous coal.
[b]Assuming 3.5% sulphur content, of which 15% remains in ash.
[c]Assuming 9% ash content and 97.5% fly ash removal efficiency.
[d]Represents the burning of about 1.7 10^5 m^3/year of oil with an assumed ash content of 0.05% and 1.6% sulphur content by weight. (Present sulphur values are closer to an average of about 0.7% .) The emissions given here assume the operation is carried out with no pollution control.
Source: Ref. 2.

Table 65: *Metals in fuel µg/g (ppm)*

	Range of values	
	Coal	Oil
As (c) Arsenic	1–15	0.02–0.06
Cd (t) Cadmium	0.01–2	< 0.01
Co Cobalt	1–40	0.2–0.5
Cr (c) Chromium	2–50	0.02–0.09
Cu (t) Copper	0.4–40	0.1–0.3
Hg (t) Mercury	0.07–1.5	0.002
Mn Manganese	25–100	0.03–0.1
Mo Molybdenum	7–20	0.07–0.3
Ni (c) Nickel	4–60	6–23
Pb (t) Lead	2–30	0.07–2
V Vanadium	5–35	11–90
Zn Zinc	5–100	0.3–1
Uranium (c)	0.4–3.7	
Thorium (c)	0.3–3.6	

(c) carcinogenic
(t) toxic
Source: Ref. 3.

are not the only, or even a major, source of some of these pollutants. For the UK the motor car discharges into the atmosphere an estimated 9,000 tonnes of lead (which is deliberately added to petrol to improve performance) compared with only some 36 tonnes from coal-fired power stations.

For the nuclear power station operating under normal conditions the hazards to the population come from controlled releases of gaseous fission products discharged to the atmosphere and from much smaller amounts of fision products discharged in liquid effluents. In the second case the effluents are treated by chemical or ion-exchange or held in delay tanks to allow the radioactivity to decay to an acceptable level before discharge. The reprocessing of irradiated fuel also gives rise to both liquid and gaseous wastes, the discharge of which is also governed by strict controls.

In most countries maximum limits are set by national regulatory authorities for the amount of activity that can be discharged. These limits generally follow the recommendations of the International Commission of Radiological Protection (ICRP) and the United Nationals Scientific Committee on the Effects of Atomic Radiation (UNSCEAR). The ICRP consists of a group of twelve scientists, who are selected by the International Congress of Radiology. The Commission is independent of all governments and political bodies and is accountable only to the International Congress of Radiology. UNSCEAR was set up by the United Nations in 1955 and reports to the IAEA.

For almost all operating nuclear stations, experience shows that the actual discharges are only a small fraction (about 1%) of the authorised maximum values. If it is assumed that the number of nuclear power stations increases in the future until the installed electrical capacity reaches 1 kW per person it has been estimated that, taking account of all stages of the fuel cycle, the average general population exposure to ionising radiation may increase to 1.5 millirem per year. If occupational exposure of staff in the nuclear power stations and at reprocessing plants is also included the overall average would rise to 6 millirem per year. This is no more than 6% of the average natural background, or 4% of the exposure from all sources including medical X-rays (Figure 48).[4]

It is only about 3% of the increased radiation dose that would be incurred by living at a height of 2,500 above sea level.

Another way of assessing the hazard of nuclear power generation is in terms of collective dose; the individual dose multiplied by the number of people within a given area. For a nuclear station one estimate puts the average collective dose from air-borne activity to population living within a 80 km radius of a 1,000 MWe station at 10–40 man rem/year, depending on the design of the nuclear reactor and also, of course, on the number of people within the designated area. This can be compared with

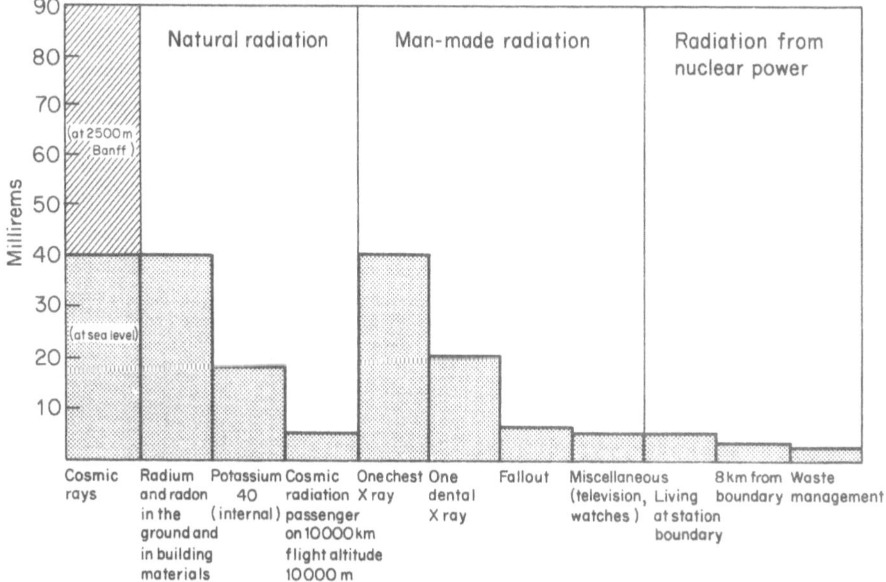

Figure 48: *Average radiation doses from various sources*

the radioactivity discharged from a coal fired power station.

Coal contains small quantities of both uranium and thorium, on average about 1 and 2 ppm respectively for United States coals, although individual samples have contained up to 25 times more. A report by the Oak Ridge National Laboratory of the US assessed the radiation exposure from burning coal (at the average values of 1 ppm U and 2 ppm Th) assuming that all the radon together with 1% of the ash is released to the atmosphere. Dose calculations were carried out for an assumed population of 3.5 million within a radius of 80 km taking account of meteorological dispersion, intake through inhalation and ingestion through food chains. The results are summarised in Table 66.[5]

This shows the whole body dose is always greater for a coal fired plant than for a PWR and that except for the very high chimney stacks the population dose is markedly higher for the coal fired plant. (But for a BWR which has a larger gaseous release the population dose would be about 20 man rem/year, and about 40 for a gas cooled reactor.) The coal figures are low estimates since the fly ash release is on average closer to 8% than the 1% assumed so the dose figures could be increased by the same factor. Although the actual radioactivity release, in terms of curies/year is greater for a nuclear plant (Table 67).[6]

Table 66: *Comparative dose commitments for coal-fired and nuclear 100 MW(e) generating plant from airborne releases*

Organ	Maximum individual dose,[a] mrem/year		Population dose,[c] man rem/year				
			Coal-fired[b]				
	Coal-fired plant[b]	Nuclear plant (PWR)	Stack height 160ft	Stack height 325ft	Stack height 650ft	Stack height 975ft	PWR
Whole body	1.9	1.8	23	21	19	18	13
Bone	18.2	2.7	249	225	192	180	20
Lungs	1.9	1.2	34	29	23	21	9
Thyroid	1.9	3.8	23	21	19	18	12
Kidneys	3.4	1.3	55	50	43	41	9
Liver	2.4	1.3	32	29	26	25	10
Spleen	2.7	1.1	37	34	31	29	8

[a]At the plant boundary, assumed to be 500 m from the point of release.
[b]Assuming an ash release of 1% and coal containing 1 ppm uranium and 2 ppm thorium.
[c]Out to 55 miles for a mid-western site assumed to have a population of 3.5 million within that radius.
Source: Ref. 5.

The heavy nuclides, radium-226 and 228 as well as thorium, polonium and actinium from the coal station have a greater biological significance than the more weakly active noble gases from the nuclear plant.

There is however great difficulty in relating the observed emissions from fossil and nuclear power stations to actual effects on the health of the public. With the much greater attention now being paid to a clean environment, measurements are now made in many countries, mapping the average levels of pollutants in and around major centres of population. The difficulty lies in quantifying the risk with the exposure level. Epidemiological evidence from major episodes, e.g. the London smog of 1952, has clearly shown that severe air pollution can cause illness and premature death. This particular event was estimated to have caused 3,900 premature deaths. It was however largely brought about by burning coal in domestic hearths, not at power stations. Such domestic consumption of coal in urban areas has now been banned in the UK under the Clean Air Act of 1956. It is also recognised that in general the occurrence of lung cancer is several times more frequent in urban than in rural areas, but the factor is variable, and there are other features, such as diet, smoking, or specific industrial exposure which complicate the establishment of a direct relationship between lung cancer and air quality. The problem

Table 67: *Effluents from 1,000 MW(e) electric power stations*

	Type of fuel			
	Coal	*Oil*	*Gas*	*Nuclear*
Annual fuel consumption	2.3×10^6 tons	9×10^6 [d] barrels	64×10^9 [d] ft^3	2500 lb[a]
Annual release of pollutants (millions of pounds)				
Oxides of sulphur	306	116	0.03	0
Oxides of nitrogen	46	48	27	0
Carbon monoxide	1.15	0.02		0
Hydrocarbons	0.46	1.47		0
Aldehydes	0.12	0.26	0.07	0
Fly ash (97.5% removed)	9.9	1.6	1.0	0
Annual release of nuclides, Ci				
1620-year ^{226}Ra	0.0172	0.00015		0
5.7-year ^{228}Ra	0.0108	0.00035		0
10.8-year ^{85}Kr + 5.3-day ^{133}Xe	0	0	0	
Radioactive noble gases[b]				
PWR[c]				600
BWR[c]				1.11×10^6
^{131}I	0	0	0	
PWR[c]				0
BWR[c]				0.85

[a]From a fuel reserve of approximately 27,500 tons.
[b]For a PWR with greater than 1 month gas holdup, these gases would be 10.8-year ^{85}Kr and 5.3-day ^{133}Xe. The typical 30 min holdup and diffusion mixture from a BWR is composed primarily of 1.3-hr ^{87}Kr, 2.8-hr ^{88}Kr, 9.2-hr ^{175}Xe, and 17-min ^{178}Xe.
[c]Calculated from average of releases during 1969, yearly totals estimated for those plants with less than 9 months of full-power available.
[d]Revised figures.
Source: Ref. 6.

becomes more complicated at the low levels of exposure over long periods of time which are now being considered. While animal experiments can give some guidance the results may not be applicable to man because of biological and species differences. Most information is therefore based on extrapolating the results from a small number of severe cases of exposure down to lower levels, where any actual effect may not be directly observable but only inferred on a statistical basis after a prolonged exposure period. The effects of radiation on man are known principally from the survivors of the atomic bombs, or from those patients who have been subjected to large radiation doses for medical treatment. But there are then difficulties in extrapolating these effects to what might be expected for much lower exposure levels (Figure 49).[7]

This is now a matter of controversy. The cautious, generally accepted view has been that the risk of cancer is directly

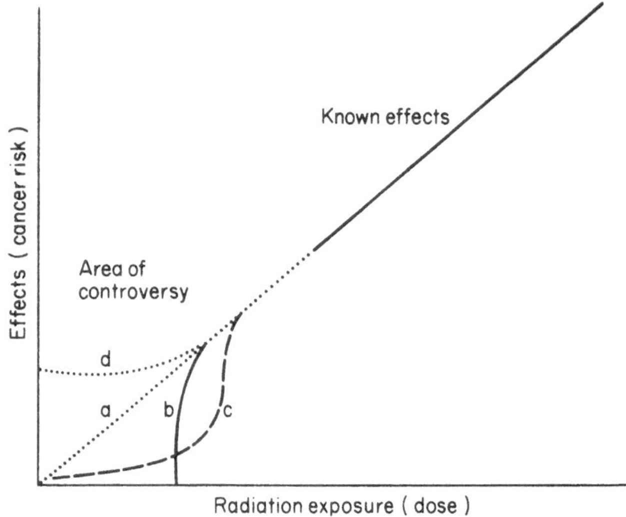

Figure 49: *Extrapolation of known radiation effects down to small doses*

proportional to dose even at low levels "a". Some would contend that there is a threshold below which the risk is essentially zero "b"; others, that the risks are lower than expected by linearity "c" or, as has been recently argued, higher "d". This is a rather sterile controversy since there seems little likelihood that, in the short term at least, it will ever be resolved. It is well nigh impossible to detect small variations in cancer rate that may be caused by radiation against the background of effects from many other carcinogenic agents, including cigarette smoking. Recent work, notably on more detailed study of the Japanese survivors, which has separated out the effects of gamma radiation of the Nagasaki bomb from the neutron radiation of the Hiroshima bomb, has shown that the risk of cancer from the gamma radiation is at least half of that expected from linearity, and since gamma radiation would be predominantly associated with nuclear power it seems that the Nagasaki experience would be more appropriate to estimate the risks from nuclear power plant. But even for this study the statistical base is small. There have been, in the 30 year study of the Japanese survivors only 77 excess leukaemia cases and about 130 other cancers among more than 100,000 persons, who as a group had been exposed to almost 2,000,000 man-rem. And this very heavy irradiation occurred in a very short period of time. It is however to be expected that radiation – as for most other factors to which the human body responds (e.g. alcohol or sunburn) will be much less damaging if the total exposure or dose is experienced slowly over a long period of time. From this point of view the linear hypothesis (curve "a") could be recognised as being over-cautious since it assumes that the number of cancers caused would be the same for a population of 1,000 people receiving a radiation dose of 100 rem in a few minutes or hours (for example the atomic bomb victims on which the extrapolation is based) as for a dose of 0.1 rem received by 1 million people over the course of one year.

It is also considered possible that at very low doses the natural repair mechanisms may become proportionately more effective. Damage induced in DNA by radiation or chemical carcinogens may be rectified at least in part by cellular DNA repair mechanisms. The dose–response relationship may then be influenced by the nature and extent of the repair process. It is interesting to note that some small groups of individuals have been identified who are specially sensitive to environmental carcinogens because of defective DNA repair ability. Another factor taken into account in setting pollution standards is the natural background level. The natural radioactivity of rocks and radiation from cosmic rays gives a level of exposure to which man is inevitably subjected. If the additional man-made contribution to the level of radioactivity is quantitatively and qualitatively only a small fraction of the natural background then the damage caused can be considered as tolerable. Figure 50 compares the natural background concentrations with actual man-made exposures in relation to the levels set by the regulatory standards (in this case the US Environmental Protection Agency) and the levels at which medically perceived effects occur.[1] It is apparent that the actual emissions of sulphur and nitrogen oxides are well above the natural background and very much closer to the level for medically perceivable effects than is the case for radiation from nuclear background.

A joint report in 1975 by the US National Academy of Sciences, National Academy of Engineering and National Research Council attempted to estimate the health effects associated with sulphur oxide emissions from power plant. The figures in Table 68 are based on a 620 MW(e) plant emitting 96.5×10^6 pounds of sulphur per year.[8]

It is however emphasised that these numerical estimates are controversial, relying as they do on limited information and arbitrary assumptions and cannot be regarded as proven results. They could be low by a factor of 2 or high by a factor of 10.

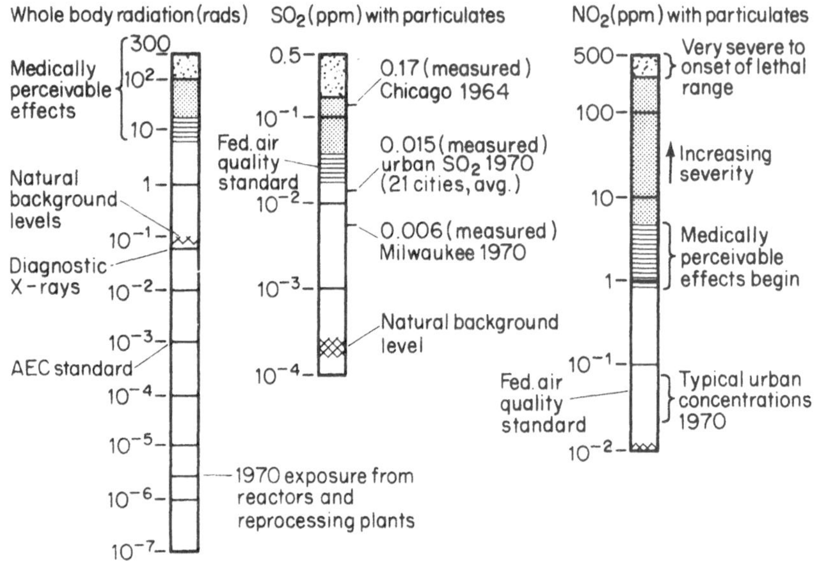

Figure 50: *Comparison of pollutant standards, background levels, man-made exposures and health effects. (Note: neither the units nor factors of 10 on the scales are the same.) (Source: Ref. 1)*

Table 68: *Health effects associated with sulphur oxide emissions*[a]

	Remote location	*Urban location*
Cases of chronic respiratory disease	25,600	75,000
Person-days of aggravated heart-lung disease symptoms	265,000	755,000
Asthma attacks	53,000	156,000
Cases of children's respiratory disease	6,200	18,400
Premature deaths	14	42

[a]Source: Illustrative calculations based on distributive models, postulated conversions of SO_2 to SO_4, and EPA epidemiological data for representative power plants in the North-east USA emitting 96.5 x 10^6 pounds of sulphur per year – equivalent to a 620 MW(e) plant.
Source: Ref. 8.

The economic aspect of the non-lethal effects should also be taken into account. Material damage is an important factor and here the effects of sulphur dioxide and sulphate levels are easier to substantiate experimentally. One estimate of the national costs of air pollution for the US for 1968, also including health effects is given

Table 69: *Estimates of the costs in the USA of damage caused by discharges of fossil fuel pollutants in 1968 ($ billion)*

Effects (loss category)	SO_x	*Part.*	*Oxidant*	NO_x	*Total*
Residential property	2.808	2.392	–	–	5.200
Materials	2.202	0.691	1.127	0.732	4.752
Health	3.272	2.788	–	–	6.060
Vegetation	0.013	0.007	0.060	0.040	0.120
Total	8.295	5.878	1.187	0.722	16.132

SO_x = Sulphur oxides (SO_2, SO_3)
Part. = dust
Oxidant = oxidising substances such as ozone
NO_x = nitrogen oxides.
Source: Ref. 9.

in Table 69 which puts the total cost at over $16 billion/year.

Attention has recently been directed to carbon monoxide which has been linked by the US Council of Environmental Quality with angina pectoris attacks among patients suffering from this heart disease. Carcinogenic substances, which are in addition almost certainly mutagenic include Benzo(*a*)pyrene which could cause lung cancer. The conclusion of a symposium held at the Karolinska Hospital in Sweden in 1977 was that the combustion products of fossil fuels in air probably acting together with cigarette smoke have been responsible for 5–10 cases of lung cancer per 100,000 males per year in large urban areas.

Despite all the uncertainties a number of attempts have been made to compare the risks associated with electricity production. The results of these assessments have been summarised in a report by Comar and Sagan published in 1976.[8] The values quoted for occupational injuries and premature deaths (Table 70 and 71) are the lowest and highest estimates from the references on which the tables are based. The results show that nuclear power is comparable to natural gas, and that both of these show a considerable factor of safety over coal and oil.

Another set of figures (Table 72), based on a 1977 paper was given by Hamilton and Manne,[10] these have been recalculated on the total fuel cycle of a 1,000 MW(e) power plant at 75% load factor.

From the foregoing analysis it can be seen that there is strong evidence for the conclusion that nuclear power shows a marked advantage over coal and oil fired electricity generation in terms of health and safety of the workers and of the general public. A recent study, June 1978, by the American Medical Association

Table 70: *Occupational injuries per year associated with operation of 1,000 MW(e) power plant (values are lowest and highest estimates from original references)*

Occupational injuries	*Coal*	*Oil*	*Natural gas*	*Nuclear*
Extraction				
Accident	22–49	7.5–21	2.5–21	1.8–10
Disease	0.6–48	–	–	–
Transport				
Accident	0.33–23	1.1–9	1.2–1.3	0.045–0.14
Processing				
Accident	2.6–3	3–62	0.05–0.56	0.6–1.5
Conversion				
Accident	0.9–1.5	0.6–1.5	0.6–1.5	1.3
Totals				
Accident	26–77	12–94	4–24	4–13
Disease	0.6–48	–	–	–

Source: Ref. 8.

Table 71: *Premature deaths per year associated with operation of a 1,000 MW(e) power plant (values are lowest and highest estimates from original references)*[a]

	Coal	*Oil*	*Natural gas*	*Nuclear*
Occupational				
Extraction				
Accident	0.45–0.99	0.06–0.21	0.021–0.21	0.05–0.2
Disease	0–3.5	–	–	0.002–0.1
Transport				
Accident	0.055–0.4	0.03–0.1	0.02–0.024	0.002
Processing				
Accident	0.02–0.04	0.04–1	0.006–0.01	0.003–0.2
Disease	–	–	–	0.013–0.33
Conversion				
Accident	0.01–0.03	0.01–0.037	0.01–0.037	0.01
Disease	–	–	–	0.024
Subtotals				
Accident	0.54–1.5	0.14–1.3	0.057–0.28	0.065–0.41
Disease	0–3.5	–	–	0.039–0.45
Total	0.54–5.0	0.14–1.3	0.057–0.28	0.10–0.86
General Public				
Transport	0.55–1.3	–	–	–
Processing	1–10	–	–	–
Conversion	0.067–100	1–100	–	0.01–0.16[b]
Total	1.6–111	1–100	–	0.01–0.16
Total Occupational and Public	2–116	1.1–101	0.057–0.28	0.11–1.0

[a]Note: Dashes indicate no data found; effects, if any, are presumably too low to be observed; and no theoretical basis for prediction.
[b]For processing and conversion.
Source: Ref. 8.

Table 72: *Deaths and disabilities due to a 1,000 MW(e) power plant at 75% load*

	Estimated deaths	*Estimated disabilities*
Coal	7–130	200–300
Oil	2–100	100–200
Gas	0.1	10
Nuclear	1–2	5–20

combined both the occupational hazard and the public hazard and concluded that "coal has a much greater adverse impact on health than does nuclear power production" and estimated that each coal fired power station results in an annual death rate 400 times that attributable to an equivalent nuclear power station.

This discussion of the air pollution effects of different means of electricity generation has concentrated on the relative effects of nuclear power compared with coal and oil; it does not mean that any of the methods have yet led to intolerable levels of pollution, bearing in mind the essential need of society for electricity and energy. In any case electricity production is only a minor contributor (less than 10%) to the total air pollution as is shown in Table 73 giving the quantities of pollutants released into the atmosphere in Sweden from different sources in 1975.[11]

Table 73: *Releases into the atmosphere 1975 (1,000 tons)*

	SO_x Sulphur oxides	*NO_x Nitrogen oxides*	*HC Hydro-carbons*	*CO Carbon monoxide*	*Dust*
Combustion					
Electricity	66	18	0.5	1	1.9
District heating	54	14	0.6	1	2.3
Domestic hearths	160	31	3.2	4	4.4
Waste burning	1.1	0.8	?	?	0.9
Combustion total	280	64	4	6	10.0
Transport	19	200	250	1370	19
Industries	390	50	140	18	140
Other	–	–	37	–	–
Total	690	310	430	1400	170

Source: Ref. 11.

One purpose of drawing attention to the lower amount of pollution from nuclear power, compared with fossil fired power stations is to show where society should concentrate the not unlimited resources that are available for improving the environment. In the electricity production sector greater attention should be paid to reducing the emissions from coal and oil fired plants rather than with an exaggerated concern over the health effects of the cleaner nuclear plants. There must in any case be a substantial increase in the amount of coal-fired plant; it will not be possible to replace this in the short or even the medium term by nuclear power. Careful attention should however be paid to the increase in the carbon dioxide content of the atmosphere as this may, by the middle of the 21st century, set a limit to the burning of fossil fuels.

REFERENCES

1. Wash-1224, *Comparative Risk—Cost—Benefit Study of Alternative Sources of Electrical Energy*, USAEC, December 1974.
2. *Health Implications of Nuclear Power Production*, WHO, Copenhagen, 1977.
3. F. Larsson, Statens Naturvardsverk, ELMIA AB Symposium, Stockholm, March 1979.
4. *A Race Against Time*, Royal Commission on Electric Power Planning, Ontario.
5. *British Nuclear Energy Society Journal*, January 1978.
6. A.P. Hull, *Nuclear Safety*, 12 May/June 1971.
7. L. Sagan, *EPRI Journal*, September 1979.
8. Comar and L. Sagan, *Annual Review of Energy*, Vol. 1, 1976.
9. L. Barret and T. Waddell, *Cost of air pollution damage*, US Environmental Protection Agency, February 1973.
10. Hamilton and Manne, IAEA-CN-36/448, Salzburg Conference, May 1977.
11. B. Hanell, Statens Naturvardsverk, ELMIA AB Symposium, Stockholm, March 1979.

12
Carbon dioxide and the greenhouse effect

The large releases of carbon dioxide into the atmosphere from the burning of fossil fuels may present a special problem. Carbon dioxide is not a pollutant in the normal sense of the word. It is an essential constituent of the environment and plays a key role in plant growth being converted by photosynthesis in plants to hydrocarbons. The burning of fossil fuel returns to the atmosphere carbon which was fixed by plants in primeval times, when the CO_2 concentration in the atmosphere was higher, about 450 ppm compared with about 325 ppm at present. The concern arises in that the rapid increase in the rate of burning of fossil fuels has produced CO_2 at a greater rate than can be absorbed under equilibrium conditions in the oceans and by plants and as a result the CO_2 concentration in the atmosphere has shown a steady rise over recent years (Figure 51).[1]

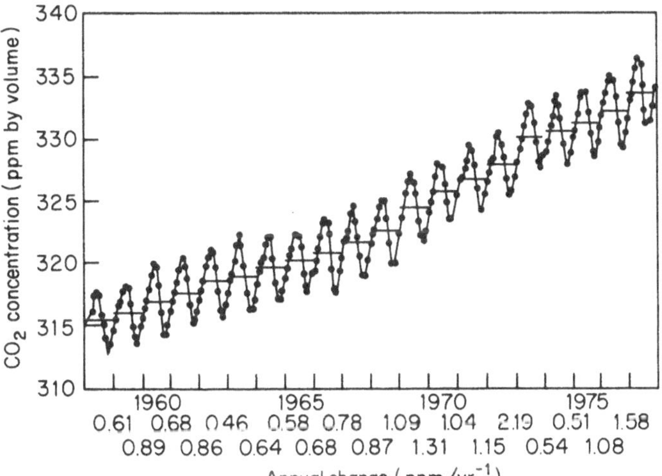

Figure 51: *Annual change in CO_2 concentration in the atmosphere. This record by NOAA's Mauna Loa Observatory, Hawaii shows monthly averages. Seasonal variations result when CO_2 is removed during the growing season in the northern hemisphere and then released in autumn and winter. The graph shows a rise of more than 5% since 1958. Current rate of increase is 1 ppm per year (2.3×10^{15} g of C). (Source: Scripps Inst. of Oceanography)*

The measured increase shows a steady rise in CO_2 concentration from 315 ppm in 1958 to 334 ppm in 1977. The level in pre-industrial 1860 has been estimated at 293 ppm. Over one-quarter of the 41 ppm increase since 1860 has come in the last ten years. It is believed that if the present growth rate in the use of fossil fuels continues to increase at 4% per year the CO_2 level could double within 30–50 years.[1] The effect of this would be to increase the surface temperature of the world. Carbon dioxide is transparent to the incoming short-wave solar radiation which passes through the atmosphere and warms the earth's surface. Part of this heat is then reflected from the earth into the atmosphere as longer wave infrared radiation. This heat is however absorbed by the CO_2 giving a warmer atmosphere and a re-reflection of heat back to the earth thus raising the temperature of the earth's surface. Glass performs in an analogous manner in a greenhouse – hence the term "greenhouse effect".

World climate model studies now suggest that if the atmospheric concentration of carbon dioxide were to double, the mean temperature of the lower levels of the atmosphere would increase by between 1.9 and 2.9°C, and the temperature of the earth's surface layers in high latitudes would increase by two or three times as much. There is then a growing consensus that increasing CO_2 levels could bring about major changes in climate. These changes could in turn bring about balancing measures such as increased plant growth and a greater rate of photosynthesis, on the other hand some oceanographers believe that the capacity of surface ocean waters to accept CO_2 will decrease so that the atmospheric concentration will increase.

The consequence of any climate change may also be regarded differently in different parts of the world, some regions may consider themselves to be favourably affected others adversely, but there is no means of predicting who will be the winners or who the losers. The most feared result is that the increase in temperature at high latitudes could lead to a melting of the West Antarctic Ice Sheets. This would in turn raise the level of the oceans perhaps by 5–18 m with the inundation of many major cities now at sea level. These effects will however occur relatively slowly, and continued monitoring of CO_2 levels, combined with increasingly accurate climate model studies should give an early warning and time for some corrective action, which could include a greater reliance on nuclear power so as to reduce the burning of fossil fuels.

There is however a danger that if the replacement of fossil fuels is left until a clear warning trend is established it may be too late to take effective action and the consequences of the climatic change could endure for thousands of years.

REFERENCES

1. *EPRI Journal*, July/August 1978.

13
Hazard of major accidents

The fear of major accidents in electricity generating plant that could cause numerous deaths amongst the surrounding population is mostly associated with nuclear power plant. This is partly because accident assessments have been carried out and published by the nuclear industry, and since the amount of potentially dangerous material contained within a nuclear plant is large these assessments have attracted wide attention, even though the probability that such an accident might occur is very low.

Much less attention has been paid to the much greater probability of major disasters, in terms of loss of life, arising from dam failures, in hydro electric plant. There is also the danger associated with the storage of hazardous materials, notably chlorine, which is used in many steam power stations for water treatment.

DAM FAILURES

The failure of water dams is a not infrequent occurrence and such failure or overflowing must be considered as the major risk associated with the generation of electricity by water power.

Table 74 lists some of the catastrophic dam failures over the past 20 years.

Table 74: *Catastrophic dam failures*

Year	*Place*	*Fatalities*
1979	Morvi, India	> 3,000
1977	Teton, USA	9–11
1967	Koyna, India	180
1963	Vaiont, Italy	2,600–3,000
1961	Kiev (BabiYar) USSR	145
1960	Oros, Brazil	~ 1,000
1959	Frejus, France	421
1959	Vega de Tera, Spain	123–150
1959	Bhakra, India	10

CHLORINE

Chlorine is widely used in industry and water treatment. At power stations (both nuclear and conventional) quantities up to 90 tons can be stored in fixed tanks. For the worst possible disaster – the total and immediate loss of a full 90-ton tank – a highly toxic cloud of gas could extend some tens of kilometres in stable weather conditions. Table 75 lists chlorine accidents that have occurred.[1]

Table 75: *Chlorine accidents*

Year	*Place*	*Chlorine lost (tons)*	*Fatalities*
1952	Wilsum, Germany	15	7
1947	Rauma, Finland	30	19
1939	Zarnesti, Roumania	25	60
1925	St. Auban, France	25	19

SAFETY OF NUCLEAR POWER PLANT

If it is accepted that the normal operation of a nuclear power plant is less damaging to man and the environment than an equivalent fossil fuel station, the question must then be considered of the likelihood and potential hazards arising accidentally due to maloperation or failure of the equipment.

The main concern is the possibility of an accidental release into the environment of some of the radioactive fission products which are formed by the fission process in the fuel of the reactor core. The usual fuel in most reactor systems is a ceramic oxide of uranium. The fission products are retained by chemical and physical bonding within the lattice structure of this material (which has a melting point of over 2,500°C) and even the volatile species diffuse only slowly to the surface where a metal cladding – the fuel can – forms a barrier to their release. Minor faults in the fuel can, sometimes lead to the release of small quantities of fission products into the main coolant stream, adding to the radioactivity from activation products already there. Continuous stripping of the coolant to remove this activity is necessary, and some of this radioactive waste is then discharged under carefully controlled conditions as gaseous or liquid effluent. This is covered by normal operation; the discharges are subject to legislative control and are required to be kept as low as practicable. The failure of individual fuel elements can be readily detected and the faulty fuel elements replaced. The rate of such failure is normally very low, and the damaged fuel is usually taken care of at the time of scheduled fuel recharge. The only way that potentially large amounts of radioactivity could be released is if the fuel in the core were to melt. The most hazardous accident is then one which could lead to a rapid loss of coolant with a possible overheating and perhaps melting of the fuel. (It must be emphasised that the runaway nuclear reaction of the type associated with an atomic bomb is physically impossible in a nuclear power station.)

Overheating of the fuel could take place if the coolant flow were interrupted, even if the reactor were to be shut down by the control system, as some of the fission products formed in the fuel continue to generate heat because of their radioactive emissions. The heat output from this source is quite large, initially about one-tenth of the reactor output, but falls with time as the fission products decay. The temperature of the fuel would then rise rapidly with the interruption of the normal coolant flow.

To guard against the possibility of an accident which might affect the public through the release of excessive quantities of radioactive materials, the nuclear industry relies on three levels of defence. The first level addresses accident prevention through the design of the plant; the second covers manufacturing procedures including codes, standards, quality assurance, redundancy, testing and inspection; the third relies on licensing and control of plant operations.

The basic protection against the release of radioactive materials as a result of a reactor accident is in the design of the plant. The safety design of reactors includes a series of systems to prevent the overheating of the fuel and to control the possible release of radiation. For a release to occur there must be a chain of failures in the reactor and also in the systems designed to remove and contain any radioactivity released from the reactor core. For the LWRs the core is cooled by the primary cooling system which is divided into several parallel circuits. If this cooling becomes inadequate either because of a large undesired increase in reactor power or of a major break – pump failure, pipe rupture etc. – in the coolant circuit, a number of safety systems are brought into action. As a first step, the reactor is shut down. The second step is to trigger the emergency core cooling. For many accidents the emergency core cooling is not necessary but it is provided should there be an unfavourably located rupture in the main stream or feed water lines. These emergency core cooling systems have independent pumps, power and water supplies. They are also designed and constructed as two or more parallel independent circuits so that if some components fail to operate the core can still be cooled.

The possibility must, however, be considered that all these systems fail and a fuel melt occurs. Radioactivity could then be released into the containment building. This is an essentially leak tight structure provided to prevent the escape of airborne radioactivity into the outer atmosphere. Some of the radioactivity will deposit on the surfaces inside the containment; some is trapped by filters and water sprays, which wash the radioactivity from the atmosphere within the containment building and reduce the pressure by condensing the steam which will flash off from the high pressure cooling system.

In the very worst case it might be supposed that the containment structure itself might ultimately fail or be damaged by missiles created by the accident, and even though a large part of the total activity would be removed by the spray systems or deposited on the inner surface of the building, some release to the outer atmosphere could occur. The consequences of this would then depend on the conditions under which the accident occurred. These include the amount of activity released, the way it is dispersed by the prevailing weather conditions and the number of people exposed to the radiation. From a public health point of view the main hazard comes from the fission product gases, krypton, xenon and iodine-131. There could also be some release of tellurium, ruthenium and caesium, but most of the fission products would be retained within the containment. An important factor is the release height. Any event leading to a major containment breach is likely to be associated with the release of large amounts of heat which would exert a lifting effect on the radioactive cloud and this would then tend to disperse as it is carried downwind. Inhalation of fission products would be the main immediate hazard but after some time the deposition of caesium – by then much dispersed – could dominate. The most important factors in determining the consequences of any release of activity are the meteorological conditions and the actions taken to mitigate the consequences such as taking shelter or possibly evacuating the most exposed population. In the case of longer term hazards from contamination, consumption of food and drink might have to be controlled for a period of time, as for example the milk production in the neighbourhood of Windscale after the accident in 1952. Other actions could include the distribution of tablets containing stable iodine which would block the uptake by the thyroid of radioiodine.

In addition to safety in design, special attention is paid to the quality of materials used, and the manufacture of equipment for nuclear reactors. All the components of a light water reactor in the primary circuit, which are subject to high pressure, have to be manufactured to the requirements of the ASME Boiler and Pressure Vessel Codes Section III "Nuclear Power Plant Components" in the United States or an equivalent national standard in some other countries. These codes include requirements for extensive stress analysis for pressure vessels and piping, taking into account all possible modes of failure, stringent requirements for non-destructive testing at all stages of manufacture and surveillance of the quality of the materials used. Quality assurance (QA) is the provision of evidence that all requirements for quality have been met. These techniques which enabled the high standards of reliability of the space exploration programmes to be met are now applied in the nuclear industry. The QA practices define direct responsibility for every activity related to the generation of nuclear power including design, purchasing, fabrication, storing, cleaning, construction and installation, inspecting, testing, operating, repairing and modifying. It is particularly concerned with those items where divided responsibility would otherwise occur. The onus for seeing that the conditions of the QA programme are met normally fall to the plant owner or operator and as indicated above this covers the plant not only during manufacture and erection but also for the whole of its operation life.

An important aspect of quality assurance is the way in which the responsibility for quality is spread throughout the whole industry. The construction of a nuclear power station is carried out by a hierarchy of organisations with the utility owner at the top, the main contractor at the next level and then an expanding pyramid of sub-contractors. Each organisation at any one level will be playing the dual role of supplier in relation to the organisation immediately above them in the pyramid and a purchaser to those immediately below. The aim of the quality assurance arrangements is to place the responsibility squarely on the shoulders of each supplier for providing plant or materials of satisfactory quality and of demonstrating that this is being done.

The third factor in ensuring the safety of nuclear plant is through licensing and control regulations. All countries with significant nuclear power programmes have enacted legislation to control the standards of design, construction and operation of nuclear power plants. This control is normally exercised in practice by a government department or statutory body (e.g. the Nuclear Regulatory Commission in the US and the Nuclear Installations Inspectorate in the UK) through the award of a series of permits or licences on a stage by stage basis. Although the procedures vary from country to country a typical sequence might be:

(1) Site licence;
(2) Construction permit;
(3) Fuel loading permit;
(4) Operating permit.

The major assessment precedes the construction permit and takes the form of a detailed in-depth appraisal by the regulatory authority of the design, quality assurance programmes and proposed operating procedures. During construction the regulatory body would undertake inspections to audit key items such as pressure vessels, pumps, valves, control systems, containment buildings etc. to ensure compliance with requirements. These inspections are separate from and additional to the detailed inspections required in the quality assurance programme. An operating permit would finally require the completion of this compliance-audit to the satisfaction of the regulatory body, the approval of the detailed operating schedules, the acceptance by the regulatory body of a completed Safety Analysis Report, and the approval or licensing of the operating staff.

The constant effort to ensure that nuclear reactors can be operated safely is supported in many countries by substantial research programmes within special organisation set up for the purpose. These include for example the Health and Safety Branch of the UKAEA, the Institut für Reaktorsicherheit in Germany. The scope of this work varies from country to country, but in general covers such items as the study of systems and methods for accident prevention: the behaviour and properties of materials under service conditions; the removal of radioactive waste from air and water; radiation protection; environmental hygiene; the development of standards and criteria. At the international level the EEC Euratom research centre at Ispra, Italy, is to carry out a major in-pile investigation on the effects of a loss-of-coolant accident on the integrity of uranium fuel rods. This work will also be supported by the USA and Japan.

These experimental programmes are designed to provide further information so that the modes of failure and the probabilities of failure can be more accurately assessed.

ASSESSMENT OF RISK

Within the emphasis on safety of nuclear power plant it is understandable that attempts should be made to quantify the potential dangers and risks of operation. A first crude attempt was made by the USAEC in 1957 in the often quoted report WASH-740. This deliberate assessment of the upper limit of the consequences of a nuclear accident, taking at each stage the most pessimistic assumptions, led to highly alarming conclusions. A much more extensive study to obtain a more realistic evaluation of the risks and consequences of a nuclear accident was published in draft form at the end of 1974 by Professor Rasmussen for the USAEC. The study first identified events such as equipment failure, poor maintenance, human error and other possible causes of malfunction that could give rise to accidents. Fault-tree analysis (a technique developed originally in the UK for aircraft safety and successfully applied by NASA for the space programme) was used to estimate the probability of each event in the accident path and the overall probability of various types of accident evaluated. The study then went on to assess the consequences of the accident, taking account of factors such as the removal of radioactivity by deposition on surfaces inside the reactor and by washing out with sprays and retention on filters, the meteorological conditions which would affect the dispersal, as well as the population distribution and possible evacuation. The conclusions arrived at for the consequence and probability of an accident in a system of 100 power reactors are summarised in Table 76.[2]

Table 76: *Approximate values of acute illness and latent effects for 100 reactors*

Chance per year	*Consequences*				
	Acute fatalities	*Acute illness*	*Latent cancer*	*Thyroid nodules*	*Genetic effects*
1 in 170[a]	< 1.0	1.0	1.0	4	1.0
1 in 10,000[b]	70	170	450	12,000	450
1 in 100,000[c]	450	900	1,300	42,000	1,300
1 in 1,000,000[d]	1,200	2,500	2,300	75,000	2,300
1 in 10,000,000[e]	2,300	5,600	3,200	84,000	3,200

[a]This is the predicted chance of core melt considering 100 reactors.
[b]About two core melts out of 100 would produce the consequences in this row.
[c]About two core melts out of 1,000 would produce the consequences in this row.
[d]About two core melts out of 10,000 would produce the consequences in this row.
[e]About two core melts out of 100,000 would produce the consequences in this row.

A similar, but quite independent study, was carried out in Sweden in connection with proposals to site a nuclear power plant close to large centres of population to provide both heat for district heating and electricity.[3] The report (*Närförläggning av kärnkraftverk*) published in 1974 came to very similar conclusions to the Rasmussen report. It noted that acute injuries can only be encountered if very large amounts of radioactive matter are released. This is only conceivable in accidents in which a number of the reactor safety systems are put out of action. Such releases can only occur if cooling of the fuel becomes inadequate as a result of some malfunction while simultaneously leakage occurs from both the primary circulation system and the reactor containment. The probability of such an accident is put at 1 to 10 per million reactor years. But it is only in combination with highly unfavourable atmospheric dispersion conditions of low probability that even the release of large amounts of radioactivity would result in a large number of acute injuries. The probability that an accident would cause such injuries is then estimated at 0.1 to 1 per million reactor years. The report concludes that for an accident in a nuclear plant sited in an urban area, of the 36 different cases studied, in only one could the number of fatalities amount to some hundreds, in three cases they were about 10, while they were zero in all the other cases. No deaths would occur at a distance of more than 2–3 km from the plant. This indicates that even the worst accidents would not differ in extent from risks already present, and accepted, within the community.

In many cases involving radio-iodine uptake by the thyroid, the most likely cause of harm, the injury is relatively mild and the prospects for successful medical treatment are good. Even in the event of a very serious reactor accident, late effects, mainly in the form of cancer and genetic effects would occur to such a small extent that it would be impossible to distinguish them from the very much larger number of similar cases that would be normally encountered in the same population.

The difficulty of these accident assessments for the nuclear industry is that they, for the first time, pose questions that have to be answered by society. How should accidents which could have very serious consequences but are of a very low probability be evaluated? What importance should be attached to some damage to health that might be suffered by a number of people as the result of a very low probability accident? The benefits of the advantages from cleaner air, less pollution, lower cost, greater self-sufficiency have to be weighed against the exceedingly small risk which however involves the possibility of large consequences.

Outside the nuclear industry these risks tend to be ignored even though tens or even hundreds of deaths at a time among the public from industrial and other disasters do occur.

An attempt in the Rasmussen report[2] to compare the risks from 100 nuclear power stations with various natural and man-made disasters only aroused controversy. The curves in Figures 52 and 53, which show the nuclear risk as much lower than almost all other risks to which man is subjected, have been the centre of an intense debate in which they have been dismissed as a public

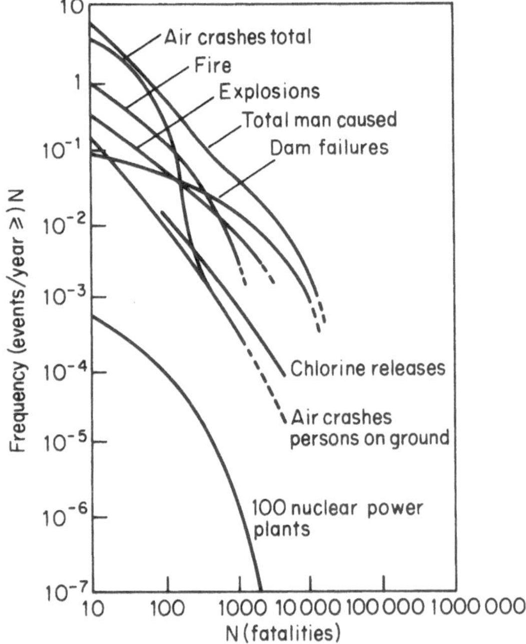

Figure 52: *Frequency of man-caused events with fatalities greater than* N

relations exercise on the part of the nuclear industry, while the main body of the report has been largely ignored. As a consequence a number of years have been wasted on sterile arguments over this particular set of figures. It would have been more sensible if the report had been accepted as a major first step towards understanding the probability, cause and consequence of nuclear accidents – which it is – and extra effort directed towards strengthening and extending the methodology and improving the accuracy of some of the estimates of equipment failure. There can be little doubt that if the Nuclear Regulatory Commission had paid proper attention to the Rasmussen Study, and extended the scope of the report to the Babcock and Wilcox PWR design, the TMI debacle would have been avoided as Rasmussen drew attention to the small loss of coolant accidents that could arise from equipment malfunction and in particular from a pressuriser valve sticking in the open position. Again on the famous bubble of hydrogen Rasmussen commented that, following a partial core melt up to 75% of the zirconium fuel cans could react with steam to give an oxide and release hydrogen, but within the reactor vessel the hydrogen–steam mixture would not be explosive.

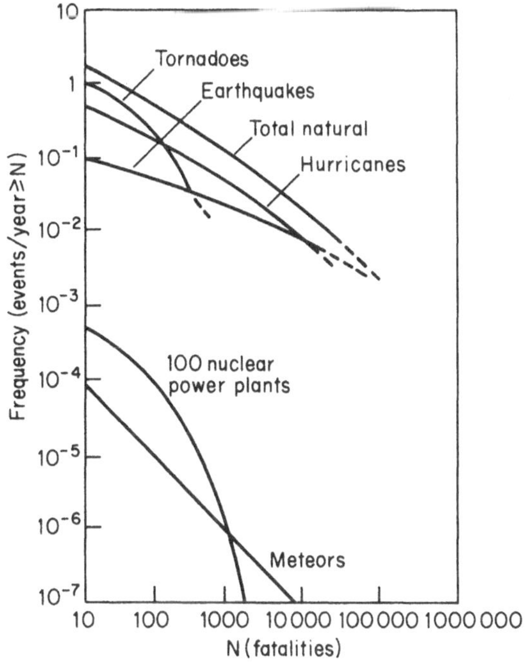

Figure 53: *Frequency of natural events with fatalities greater than* N *compared with the 100 US nuclear reactor case*

The whole philosophy of reactor safety has possibly been distorted by an early obsession with the high consequence low probability accident. This approach was partly fostered by the concept of the "maximum credible accident" as in the early WASH-740 report. But this is now being replaced by an increased understanding of risk. It is not, and never will be, possible to show that accidents cannot happen. But it is possible to estimate the probable likelihood of an accident occurring and to show that the consequences are reasonably understood and tolerable in relation to their estimated frequency. This approach – a major advance in safety philosophy – was pioneered by F.R. Farmer of the UKAEA. The emphasis for nuclear safety is on a greater understanding of the uncertainties in accident and consequence modelling to establish an increased degree of confidence that the system, its behaviour and the consequences are well understood, rather than on a prolonged description of the devices installed to minimise the chance of an accident and a demonstration of their reliability.

With increasing data from actual operating experience it will be possible to estimate the probability of accidents having minor and moderate consequences with growing accuracy. At this level considerations of operator safety and plant economy could become the overriding factor if the consequences to the environment outside the plant are small.

There is also an urgent need to apply the philosophy of accident assessment to other areas of activity, not only in the energy field but to industrial operations in general.

TOTAL RISK ACCOUNTING

The final step in the analysis of the risk of energy production was taken in a study by Dr Inhaber of the Atomic Energy Control Board of Canada, published in 1978. In this he tried to assess the total risk of different energy systems. This total risk takes account of not only the occupational and public hazards of obtaining the fuel and operating the plant hitherto considered, but also all the other risks involved, starting with the production of raw materials used to make the plant and equipment, the manufacture of this equipment, the construction of the plant, waste disposal, decommissioning, and the transport associated with all the above stages. The risks were calculated for each type of industrial activity, using the available statistical data for the deaths, injuries or time lost due to disease in producing a unit weight of material. This was then combined with an estimate of the quantity of material required. In a similar way the risks of construction were calculated from the labour statistics for the occupations taking part, giving a risk incidence per unit of time, multiplied by the time for which that occupation or trade was employed in the construction of the energy system. In calculating the public health risk, account was taken of potential catastrophic accidents such as might occur with a nuclear power station: but oil fires, natural gas explosions and dam failures were also taken into account.

For the non-conventional energy sources material acquisition and construction effort is a not inconsiderable item since, because of the lower energy density inherent with solar, wind and wave energy, these have greater material requirements per unit of energy output than for conventional power systems. Energy storage and back-up are also taken into account. Storage is required for all energy systems; in some cases it is negligible; the nuclear fuel bundles occupy only a small volume; coal stock piles and oil tank farms are much larger and the construction and storage of these attracts a greater risk. Some of the non-conventional energy sources will require energy storage or back up supplies from conventional power to cover those periods, possibly large, when their energy is not available because of seasonal or daily variations in the amount of sunshine and wind. "The sun doesn't always shine and the wind doesn't always blow, but the consumer always wants reliable heat and power." These energy sources must then carry the risk associated with the construction and operation of the back up power supply. Some of the non-conventional sources of energy may also require considerable maintenance involving replacement of materials and construction work.

Because of the many assumptions made and the uncertainties of the estimates used, the results of this total energy risk accounting can only be regarded as giving broad indications of risk levels, and this is reflected in the wide range of the risk figures put forward for each source. These expressed as total deaths (occupational and public) per 1,000 MW years of net electrical output, are given in Table 77. See also Figures 54 and 55.[4]

Table 77: *Total deaths per 1,000 MW years of net electrical output*

Coal	5–160
Oil	2–140
Gas	0.4
Nuclear	0.3–1.6
Hydro	3– 6
Wind	30– 70
Methanol	30– 70
Solar space heating	9– 10
Solar photovoltaic	20– 60
Solar thermal	8– 50

While the low position of nuclear power relative to coal and oil is not unexpected the much higher risk figures for the non-conventional sources has caused astonishment and some considerable

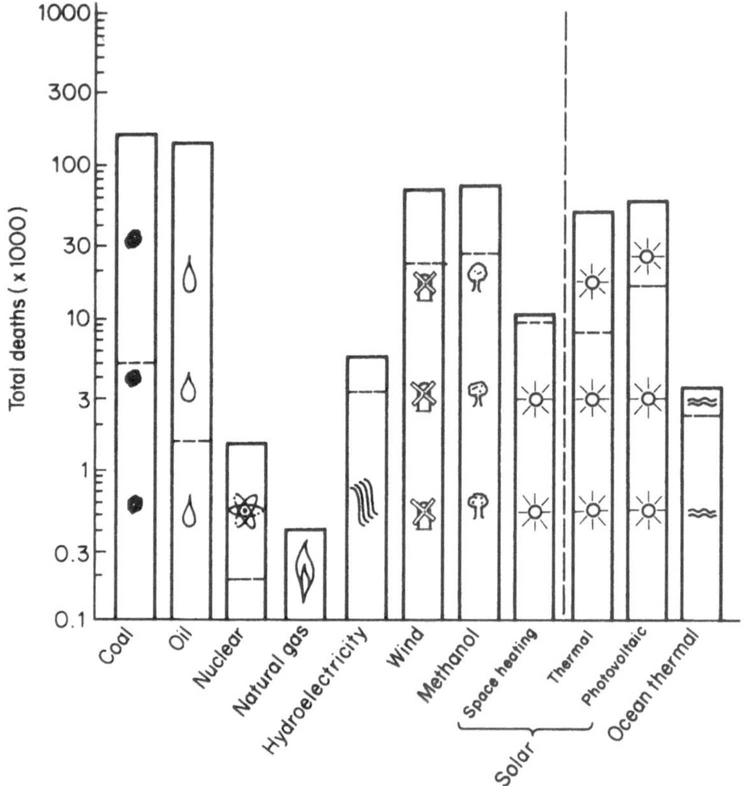

Figure 54: *Total deaths per megawatt-year by energy system. Public and occupational deaths are combined. Natural gas then nuclear power have lowest values. Risks due to non-conventional technologies are comparable to coal and oil. (Source: Ref. 4)*

resentment amongst the protagonists of the so-called "benign" energies. But as has been already noted the explanation lies in the low energy densities, the large quantity of materials and long construction time as well as the need for back up power – assumed to come from coal.

The methodology of this first attempt to assess the total risk of different forms of energy seems to be valid, and although some of the detailed estimates of accident rates, quantities of materials etc. might be questioned, every effort should be made to refine and extend this work. It is only by appreciating the real level of risk that a serious attempt can be made to reduce the hazards of all energy sources and to make a sensible allocation of the not unlimited resources that society can devote to this end.

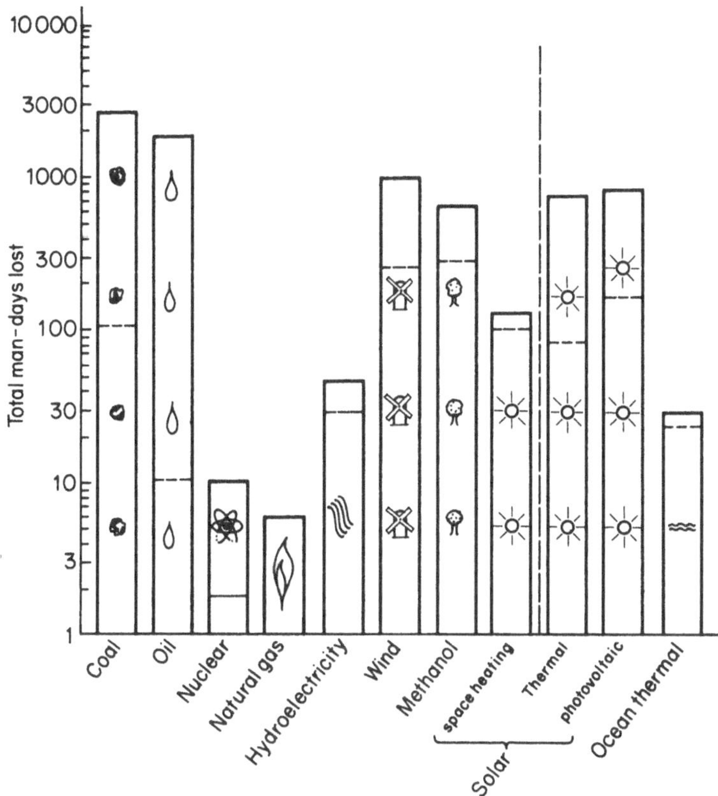

Figure 55: *Total man-days lost per megawatt-year net output over lifetime of system. The graph is similar to Figure 54 with a few differences, e.g. between lower limits of coal and oil. Natural gas and nuclear power are still lowest and non-conventional technologies are still high risk, comparable to coal and oil. (Source: Ref. 4)*

REFERENCES

1. Health and Safety Commission, *The Hazards of Conventional Sources of Energy*, HMSO, London, 1978.
2. From the Rasmussen Report, reprinted by Nuclear Engineering International.
3. Närförläggning av Kärnkraftwerk, SOU 1974:56, Industridepartment, Stockholm.
4. H. Inhaber, *Risk of Energy Production*, Atomic Energy Control Board, Ottawa, Canada, 1978.

14
Three Mile Island

No discussion on reactor accidents is complete without a reference to the accident which took place on 28 March 1979 at the TMI-2 nuclear power station near Harrisburg. Although not the first reactor accident, nor even the worst in terms of radioactivity released, TMI-2 has had a profound and shocking impact not only in the United States but around the world. It is surprising in view of the minimal releases of radioactivity that occurred that this event was given such wide publicity in all countries – in Japan, for instance, a public opinion poll has indicated that 82% of the population were aware of the accident. The responsibility for much of the hysteria, or even panic, which took place must be ascribed to the sensational reporting by the media. (The notable exception being the local press and radio which was informative without being sensational. The comment of one American observer was that "the confusion over TMI grew in direct proportion to the distance away from the island".) This arose not only from the general lack of understanding but also from the difficulty of the technicians in communicating information to non-technical journalists; from the confused and sometimes apparently contradictory reports and statements issued by the regulatory and state authorities and the utility; from the suspense that was allowed to grow over the possibility that the now notorious "bubble" of hydrogen might explode and rupture the plant – it would not, and could not – and the misunderstanding of this by the NRC which led to the somewhat confusing recommendation to evacuate pregnant women and young children from the surrounding area. All these factors were treated by press and television in the wildest terms, yet the fact that during this same period some 6,000 people were evacuated from their homes following the derailment in Florida of a train carrying toxic materials passed practically unnoticed. And when the NRC itself, about one month later issued a public statement that there was never any threat of a hydrogen explosion and admitted that it had made a mistake, this was largely ignored by the press.

Perhaps the most important lesson to be learnt from TMI is the need for a greater understanding of the psychological impact of a nuclear accident. The fear that is intensified by the fact that the five human senses cannot detect a release of radioactivity; there are no visible signs, the plant appears as normal. The population is then completely dependent on the assessment of the consequences and course of the accident by the technologists in charge of the plant or by the scientists of the regulatory authority and on the information that they give. This fear of the unknowable can only be dissipated if there is complete confidence in those who are in charge.

The conclusion of the Kemeny Commission – the President's Commission on the Three Mile Island Accident under the chairmanship of John Kemeny – was that the accident occurred as a series of human, institutional and mechanical failures.

It originated at 4 a.m. with a trip of a condensate pump which led immediately to a feedwater pump trip. This resulted in turbine and generator trips. After eight seconds the reactor was automatically shut down following a rise in pressure. At this stage one of the contributory factors to the accident occurred when an electromagnetic relief valve opened – as it should on the high pressure signal – but failed to close as the pressure dropped. The continuing loss of coolant through this valve brought in, again as it should on a low pressure signal, the emergency core cooling system which injected water into the core, some two minutes into the accident. A second error occurred when, in response to an indication of a high water level in the pressuriser, the ECCS system was manually cut back to an insufficient flow. At this stage it was discovered that the valves of the emergency feed water pump were shut, they were then only opened some eight minutes into the accident -- although an obvious maloperation it is not certain that this caused any severe damage by itself.

Another point, which had no direct bearing on the course of the accident, but gave rise to public alarm, was the automatic start up of the containment sump pumps, which pumped the slightly radioactive water held inside the containment building to an auxiliary building outside the containment. It was the discharge of a small amount of radioactivity from the ventilation system of the auxiliary building which initially gave rise to concern for the safety of the surrounding population. This was an error in design, due to the current obsession with hypothetical very large accidents; so far this was only a minor accident and the automatic containment isolation which would have been activated when the pressure in the containment building exceeded 4 psig did not observe any significant increase in radiation level, pressure, or temperature, and so did not operate. This allowed the slightly active water to be pumped out of the containment building. These sump pumps which started about eight to ten minutes into the accident were turned off after 38 minutes, so that this initial discharge of radioactivity from the containment building lasted only one half hour.

After this point the event, or rather the reasons for certain actions being taken, becomes confused. It does, however, seem that in their concern to minimise the effects of the accident, and to prevent damage to the plant, a number of actions were taken which, with hindsight can be seen to have worsened rather than improved matters. These included shutting off the main circulating pumps and reopening the electromatic relief valve which had eventually been closed. It has been suggested that the operators were over-concerned with the instruction – under normal operating conditions – to avoid completely filling the reactor circuit with water. As a consequence it seems that the reactor core was uncovered on perhaps three separate occasions for periods up to several hours. As a result of this there was serious damage to the core. It has been estimated that about 40% of the zircaloy cladding of the fuel reacted with the oxygen in the water, liberating hydrogen – hence the formation of the hydrogen bubble – but there was never any possibility of this exploding inside the reactor vessel as the oxygen from the water was taken up by the zirconium to form an oxide. There was however evidence of a small hydrogen explosion within the containment building, but this was well within the limits of the design pressure. As a result of the oxidation, about 35% of the core inventory of noble radioactive gases was released. But the most striking feature is that there was no evidence of any melting of the fuel. This is in some ways encouraging, since it has previously been assumed in most safety assessment models that significant core melting would occur if the core were uncovered for only a few minutes. This fact, combined with the way in which all the automatic safety systems operated as designed and the fact that the greater part of the radioactivity was safely contained inside the containment building, does show that a nuclear reactor can withstand a serious accident and some maloperation without any damage or harm to the environment or surrounding population. Indeed one inquiry has concluded that:[1]

> ". . . the emergency procedures were adequate to have prevented the serious consequences of the accident if they had been permitted to function or be carried out as planned."

Although this was a serious accident in terms of damage to the plant and of loss of output, the real consequences outside the plant were minimal.

In a radius of 20 km around the plant the radio iodine content of samples of milk taken from a number of different farms was in most cases too low to be measured. A special analysis showed that the iodide level was between 10 and 40 picocuries compared with 12,000 picocuries before regulations prohibit the open grazing of the animals.

The Kemeny Commission estimated that between 28 March and 15 April 1979 the collective radiation dose received as a result of the accident by the population living within a 50-mile radius of the plant was approximately 2,000 man-rem, but this can be compared with the annual dose received by the same population from natural radiation of about 240,000 man-rem. The additional dose due to the accident is less than 1% of the natural background. The Commission concluded that the radiation doses received by the general population as a result of the accident were so small that on the basis of present scientific understanding of radiation effects there will be no detectable additional cases of cancer, developmental abnormalities or genetic ill health. The major health effect of the accident is said to be mental distress produced among certain groups living within 20 miles of the plant.

The picture which emerges is of confusion amongst the regulatory officials and operating staff and of hysteria in the press, which turned what need only have been a relatively minor accident into a major national trauma. Yet despite the errors made it should be noted that the plant safety installations worked as intended and that, apart from the minimal release, all the radioactivity was safely retained within the reactor containment building: a striking demonstration of the designed safety of nuclear plant. There is however clearly a need for fundamental changes in the United States in the organisation, procedures and practices, and attitudes of the Nuclear Regulatory Commission and of the nuclear industry, and these are spelled out in detail by the Kemeny Commission.

Although this event at Three Mile Island has seriously harmed the public perception of nuclear power, it has provided lessons which have been studied by nuclear regulatory bodies, electrical utility companies and the nuclear industry throughout the world. One of the few positive conclusions to be drawn is the way in which the US authorities and the company readily made available all the information on the accident and received at the plant itself a large number of delegations from many different countries studying the cause and effect of the accident. As a result of this, small alterations to the instrumentation of a number of PWR reactors have been made in several countries, the training and qualification of plant operators is being reviewed, and more attention is now being given to minimising the consequences of minor accidents which must be expected to occur at relatively greater frequency.

The most serious real consequence of the accident is the financial harm to the investor owned utility company, General Public Utilities Corporation, which now has to face the monetary burden associated with the $700 million investment in the plant which is now shut down; these costs could amount to $100 million for each year the plant is out of service. There is in addition the cost of replacement power estimated at about $20 million per month. The Kemeny Commission estimates the total cost of the accident, including the plant clean up and a part of the waste disposal cost, at between $1,000 and $1,860 million. Utility companies now look more seriously at the financial risks associated with the construction and operation of nuclear plant and consider how these risks might be minimised. Possibilities being discussed include the joint ownership of nuclear stations – 6 utilities taking a one-sixth share in six stations rather than owning one each – and in making pooling arrangements for replacement power in the event of a serious plant failure. GPU have pointed out that prior to 28 March 1979 their nuclear plants had produced customer benefits of close to $800 millions, which went directly to their customers through the application of a fuel adjustment clause for the lower cost of nuclear fuel compared with coal and oil.[2] Some more equitable means of sharing the financial risks and benefits should be worked out.

It must finally be concluded that the major cost to the United States is in the damage TMI has done to the leadership which the US formerly held of the worlds nuclear industry. As the *Financial Times* has commented:[3]

> " . . . many of the 50-odd nations with a serious interest in nuclear energy today are already prepared to write off the US, in many cases (outside the Eastern Bloc) their leader and mentor for the past 25 years. This is only partly because they see the US as having let them down by allowing the Three Mile Island reactor to run amok. Rather they see the accident as the last straw, at the end of a decade in which the US exported world-wide its opposition to nuclear energy, and then tried to force the world to adopt ill-prepared theories for dealing with the problem of nuclear weapon proliferation by methods which were plainly more advantageous to the US than other countries.
>
> As a result, the other nations are now preparing to proceed with nuclear energy independently of the US. They will look to the US neither for technological leadership nor as a major nuclear supplier. They believe they need nuclear energy too urgently for that. France, for example, expects to sever its main

remaining link with the US in 1982 and may well by then have become the acknowledged world leader, unless West Germany very quickly regains the nuclear momentum of the early 1970s."

REFERENCES

1. NRC Office of Inspection and Enforcement, *Nuclear News*, 32 (September 1979).
2. B.H. Cherry, "The role of nuclear power in the US in the aftermath of TMI", Uranium Institute Conference, September 1979.
3. *Financial Times*, 23 October 1979.

15
Waste disposal

The so-called problem of radioactive waste, its treatment, storage and disposal has been the subject of much public discussion and some controversy in recent years; it is sometimes put forward as a major obstacle to the use of nuclear power. The difficulties arise partly over fear of radiation; misunderstanding of terms such as half-life; and in particular over the connotations of the word "waste". In present society waste is unwanted valueless material that is rejected into the environment, sometimes giving rise to social and welfare problems, when the waste is dumped without sufficient care. Man-made mountains of slag from the mining and metal industries are to be found in all industrial areas often containing toxic materials, and these have in some instances led to disaster as at Aberfan in Wales in 1966. Other wastes have been disposed of into rivers and the sea without any concern or attempt to evaluate their impact on the environment or on future generations. The discharge of mercury containing waste into the sea at Minamata in Japan caused some disability and prenatal poisoning amongst the surrounding population.

The discussion on nuclear waste has been distorted by the fact that all radioisotopes decay; by ejecting sub-atomic particles they transform into stable elements. In some this is a highly active process over relatively short periods of time, while for others it goes on at a very slow rate in some cases for millions of years and these approximate to stable elements; any hazard arises more from their toxicity than from their weak radioactivity. The concern about the long half-life (the time taken for half of the atoms present to decay) is then irrelevant; many toxic heavy metal wastes from metal mining and extraction, cadmium, mercury etc. and lead, deliberately added to petrol and discharged into the atmosphere from motor vehicle exhausts, have an infinite half-life; these elements do not decay and their potential danger will never decrease – whereas radioactivity will eventually die away. Yet no one expresses grave concern over the burden on future generations that will be caused by the discharge of these industrial wastes. Even with the question of lead pollution, it is only the health of the present generation of children that is being discussed, but the stable elements once discharged into the environment will remain a potential hazard for ever.

With the present concentration of attention on the pollution problems of our technological society, it is sometimes overlooked that the activities from which they arise are often a large improvement over earlier methods. At the turn of the century when horses provided the motive power for urban transport the daily deposit of manure and urine on the city streets became a major problem; it was estimated that in New York City these amounted to 1,000 tons per day of manure and 250 tons per day of urine. From the point of view of reducing pollution, the internal combustion engine is a great step forward.

For the nuclear industry it has been accepted from the start that the potentially harmful fission product waste shall be managed in such a way that it will be completely isolated from the human environment while it is in a dangerous state.

As is often the case in the nuclear industry public concern has been focussed on the dramatic, rather than the real problems. And in the public mind nuclear waste is the so called high level waste (HLW), the highly radioactive fission products and actinides that are separated when spent fuel is reprocessed. Yet the methods of dealing with this are greatly simplified by the relatively small volume involved. Reprocessing the fuel from a year's operation of a 1,000 MW reactor would give rise to about 15 cubic metres of liquid HLW which could be reduced to 2–3 cubic metres of solid waste if it were combined in solid form.

The immediate and temporary solution has been to store the liquid waste in multiwalled stainless steel tanks provided with cooling circuits. There is no practical reason why, with the small volumes involved, this method of storage should not be used for some 50 or even 100 years. There are now, from reprocessing the fuel from civil nuclear power stations, only a few thousand cubic metres of liquid high level waste in USA, Europe, India and Japan. This waste is stored in tanks of improved design which have not developed any serious leaks since their introduction twenty years ago. If all the fuel from the nuclear power plants in the OECD countries up to 1990 were reprocessed this would give rise to less than 50,000 cubic metres of liquid HLW. There has, however, been mounting public alarm which regards these interim arrangements as inadequate and demands that industrial scale waste management schemes should be in full operation. In some countries there is now a requirement that there should be a demonstration of "completely safe" waste disposal arrangements as a precondition of issuing a construction licence for a nuclear plant. It would however be foolish to rush headlong into an irretrievable disposal scheme: there are a number of methods that could be used and these should all be carefully investigated. One difficulty is that the quantity of waste available is as yet almost too small to enable a full scale industrial disposal scheme to be tested and put into operation.

Any waste treatment scheme must fulfil two requirements: it must accommodate or remove the decay heat and also provide a long term containment of the waste. These requirements however arise from different sources. The decay heat results from the radioactive decay of the fission products, while the need for very long term containment is determined by the presence of actinides – those elements of atomic number greater than uranium which are formed in the fuel by the capture of neutrons. This has led to suggestions that the two fractions might be separated. It is even possible that a practical use could be found for the fission products as a source of heat or radiation, while the separated actinides could be "destroyed" by re-irradiating them in a reactor. A feasible separation process with a high decontamination factor has however not been fully established and unless a use could be found for the products it would certainly be more costly than the alternative of disposing of the waste directly. In processing the waste the first objective is that the radioactive waste elements should be concentrated into a small volume and combined in a stable form which has an extremely low solubility in water. One process on which most work has been done is to combine the waste chemically as a glass. Other processes using a synthetic rock have been proposed.

In France, at the Marcoule reprocessing plant, 12,000 kg of glass which originally contained 4 million curies of waste have been produced on a pilot scale plant. The glass is cast in cylindrical blocks 30 cm in diameter and 1½ or 3 m long which are then sealed in stainless steel cans. Fifteen of the 3 m blocks would be equivalent to the HLW from one year's operation of a 1,000 MW(e) power plant.

The next step is to dispose of the glass blocks. From the start of the development of the solidification processes, it was always intended that the solid waste should eventually be buried deep underground in a stable water-free geological formation, so that the waste cannot be leached out by water and eventually returned to man.

In choosing a disposal site one of the key criteria is then the absence of ground water, or the alternative of a non-permeable rock formation. This requirement can be met by deep salt beds which are also attractive because the salt has good heat transfer properties and is plastic, so that any cracks and fissures that may form would be self-sealing. Moreover the very existence of the salt beds indicates the absence of ground water flows over the past several million years; had such water penetration occurred the salt would have been leached away. Hard rock formations, such as granites, or other igneous rocks, would also be suitable where they can be shown to have few faults or fissures and are free from ground water intrusions. For instance the plutons, large uniform plugs of igneous rock which are found in Canada throughout the Canadian Shield have remained undisturbed since they were formed some two billion years ago. The interior of a pluton is also isolated from moving ground water.

To dispose of the vitrified waste, it would be deposited in deep holes drilled into a suitable formation. But for an initial period it is possible that the heat released from the decay of the radioactive fission products may cause cracks and fractures in the surrounding structure – although this is unlikely with a salt bed which is one reason why these have been proposed – which might allow the penetration of ground water. To minimise this heating effect the waste containers would be separated at appropriate distances from one another. The longer this burial can be delayed, the less will be the amount of radioactive material remaining, and the smaller the rate of heat release. This then enables the glass blocks to be placed closer together, so that the disposal site occupies a smaller volume.

Storage for a period of thirty years before burial would reduce the amount of heat generated by half. Longer periods of storage could be perfectly feasible given the small volume to be stored and the minimum of supervision that would be required. Figure 56 shows how the heat output of a waste-glass block would decay by a factor of 10 in just over 50 years.[1]

But in the event that ground water should ever come into contact with the vitrified waste the glass blocks will be encased before burial in some suitable material such as stainless steel. In some cases additional metal casings have been proposed. The following diagrams from a Swedish study, *Handling of Spent Nuclear Fuel and Final Storage of Vitrified High Level Reprocessing Waste*,[2] show that in this case most of the plant will be located underground. The only surface facilities are an entrance building with administration and service quarters and ventilation inlets and outlets, Figure 57.

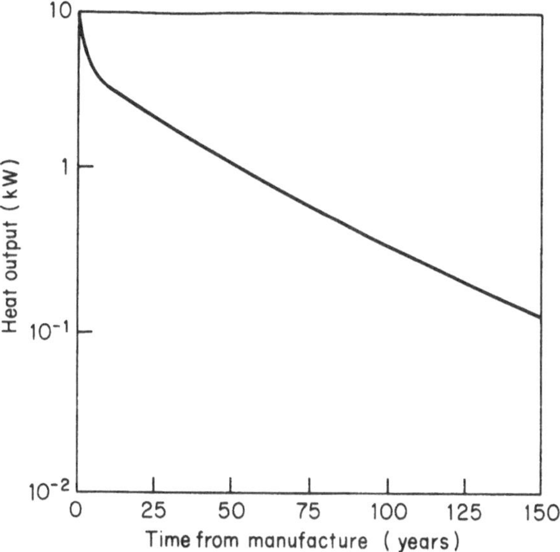

Figure 56: *Heat output from a harvest block. (PWR waste vitrified 3 years after fuel discharged from reactor.) (Source: Ref. 1)*

Waste cylinders are brought in through an access tunnel to the intermediate storage where they are placed in steel pits in concrete trenches covered by a concrete slab. The facility has four trenches in two groups with a total capacity of 6,000 waste cylinders. To cool the blocks air is circulated by a forced ventilation system, but even in the event of a total failure natural air convection would provide sufficient cooling.

At the end of the intermediate storage period the waste cylinders are transferred to the encapsulation plant where, in this proposal, they will be further enclosed in a lead-titanium canister. They are then placed deep underground in the final depository, where the canisters are placed in storage holes in the rock and packed in with a buffer material consisting of a mixture of quartz sand and bentonite, a material with low permeability and high ion exchange capacity (Figure 58).

Once placed in the disposal site the only way, apart from some very rare natural events, in which the radioactive materials could find their way back to man is through the action of ground water. The aim of the disposal concept is to delay the possibility that this might occur to such a time that the radioactivity has decayed to a harmless level. This can be ensured by providing a number of barriers each of which would have to be breached successively:

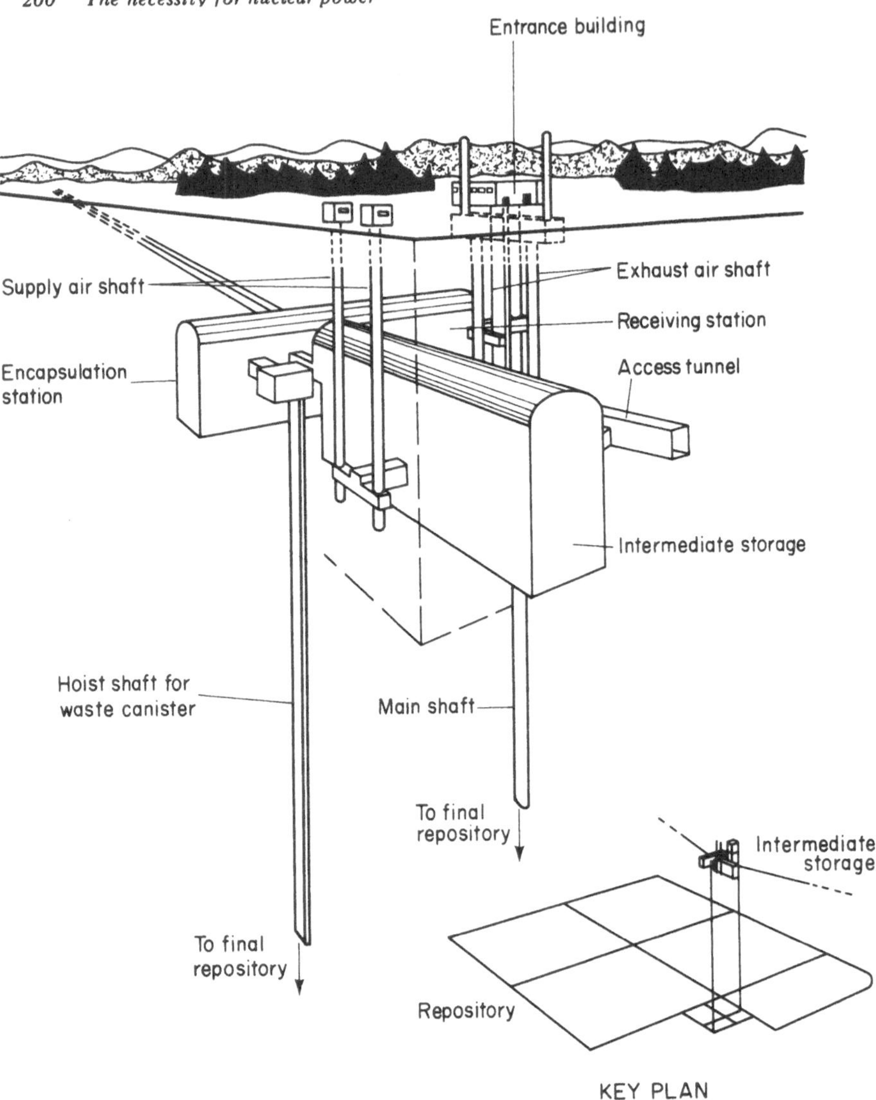

Figure 57: *Perspective drawing of plant for intermediate storage and encapsulation. It is located underground with a rock cover approximately 30 m thick. The plant is located above the final repository. (Source: Ref. 2)*

(1) The disposal site selected is one where there is clear evidence of the absence of any ground water intrusions over periods of millions of years and where no change is to be expected.

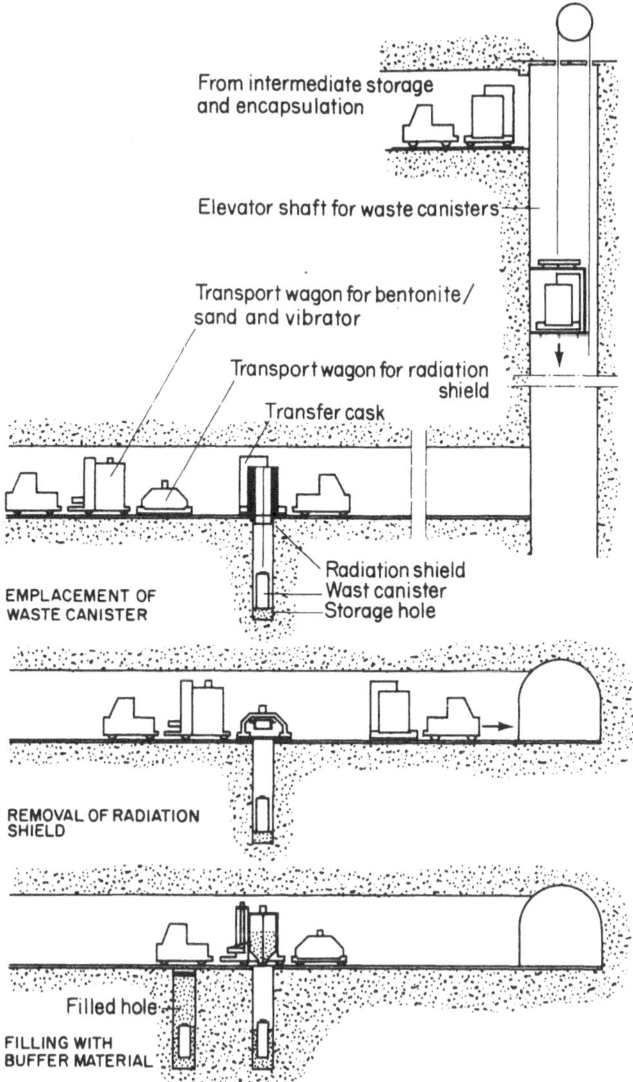

Figure 58: *Handling of waste canisters in final repository. (Source: Ref. 2)*

(2) In the event that ground water did reach the waste blocks the containers would then have to be penetrated. This could take several 1,000 years. Archaeological evidence indicates that lead and bronze can resist corrosion in sea water for more than 1,000 years.

(3) Once the container has been appreciably attacked the glass block, or other solid waste material, is then exposed. But these materials have been selected for their very low leach rate and it could conceivably take several thousand more years before an appreciable amount of the glass or waste material was dissolved.

(4) The fission products or actinides dissolved in water would then be held up by the ion exchange properties of the bentonite packing materials and by filtration and surface absorption.

(5) There is finally the long migration pathway from the deep repository to the surface. This not only adds delay but in the process the concentration of fission products would be diluted by a large factor due to mixing with other water sources.

The Swedish study assessing the waste storage proposals and pathways back to man has concluded that even with assumptions that provide a large margin of safety, the critical group of people – those who might take their drinking water from a deep well drilled in the vicinity of the final depository – could under unfavourable circumstance only be exposed to a maximum individual radiation dose of 13 millirem/year in addition to the natural background, and this additional dose would only occur after about 200,000 years. For comparison the natural background radiation in Sweden is about 100 millirem/year and the drinking water obtained from some drilled wells in Sweden gives a radiation dose of up to 30 millirem/year. This is an indication of the extraordinary degree of protection being considered, going far beyond what is required or expected for any other, and potentially equally dangerous, wastes. The rare natural events that might disturb the depository could be the impact of a meteorite, sufficiently large to disrupt the disposal site at a depth of some 500 m or the action of a new ice age where in a few areas the effect of glacier erosion might, over a very prolonged period of time, remove sufficient of the overlying rock to uncover the waste; but the timescale, the shielding effect of the glacial ice, the absence of any large population groups choosing to live in close proximity to a glacier make this too remote to be considered as a serious risk.

To become a hazard to man, the wastes buried underground would first have to be dissolved and leached out of the insoluble glass and transported by the water in which they were dissolved to the surface. Subsequently they would have to pass in sufficiently large quantities into surface waters, rivers and lakes which provided a source of drinking water directly, or be incorporated into plants or animals used as a source of food. But evidence of the leach rates which actually take place with natural radioactive materials suggests that the amounts that would reach man by this process would be insignificantly small. For instance, a naturally occurring radioactive element, radium, together with its daughter products is widely dispersed at an average concentration in soil and rock of 2.7 parts per million. This is subject to leaching and transport by ground water, but it has been estimated that only 2 parts per hundred million of the radium in the soil is so transported annually.

Further evidence of the low leach rate of radioactive fission products in the soil is given by a study of the "natural" nuclear reactor which existed in a uranium deposit at Oklo in Gabon. From an analysis of the isotopic content of the uranium it appears that about some 1,800 million years ago, when the content of the fissile isotope ^{235}U would have been higher, an intrusion of water, which acted as a neutron moderator, enabled naturally occurring fission reactions to take place over a period of about 500,000 years, during which time about 1.5 tonnes of plutonium were formed. Geochemical studies have shown that while certain nuclides have migrated from the reaction zone both the plutonium and the low solubility fission products remained fixed in the geological formation until they had further decayed to stable non-radioactive nuclides. That this should occur even in the presence of ground water is a striking example of the successful isolation of long-lived radioactive products in geological formations.

The elaborate geological disposal schemes now being worked out show the extraordinary lengths to which the industry is prepared to go to demonstrate that a "completely safe" waste disposal scheme is feasible. Indeed there is now a growing feeling that the nuclear industry has over-reacted to the demands that it must demonstrate "completely safe" disposal procedures. It can be argued that such a requirement is both unreasonable and unrealistic. It is unreasonable because the radioactivity of the wastes will cease to be significant hazard long before it has completely decayed, and it is unrealistic to demand scientific proof of safety procedures extending over a period of several hundred thousand years. A more rational approach is to determine the time for which the wastes might be considered dangerous by making a comparison with uranium in equilibrium with its radioactive decay products. This is on the reasonable assumption that a man-made hazard from waste disposal is not significant if it is no greater than that from naturally occurring uranium which is widely distributed and part of the natural environment of mankind.

When the radioactive decay curves of the specific nuclides present in the waste are examined it can be seen that the highly active fission products strontium-90 and caesium-137 which contribute most of the initial activity decay comparatively quickly and after some 500 years make only a very small contribution to the total activity. After a thousand years the activity comes from the long half-life actinide elements, americium, technetium, neptunium, plutonium and eventually radium formed as a daughter product from uranium (Figure 59).[2] The ingestion hazard to man from these elements is qualitatively very similar to that from uranium. They have also essentially the same mobility of movement when migrating with ground water through the soil. It is then possible to compare the relative toxicity of the glassified waste with a naturally occurring radioactive mineral such as pitchblende.

Figure 60 plots the fall in the relative toxicity with time of fission products and transuranic elements in glassified waste. The horizontal band indicates the ingestion hazard of the naturally occurring mineral, pitchblende.[3] This suggests that the significant period for waste disposal concern is no more than 500–1,000 years since after that time the hazard from the waste is of the same order as that from natural minerals. For such a period of time, it is possible to produce realistic scientific evidence for the behaviour of the fission product containers and the vitrified waste.

In making this comparison it should also be considered that many of the richer uranium ore deposits lie close to the surface and are being worked by open pit mining. Some natural springs, streams and other surface waters contain measurable quantities of uranium, and this is indeed one method of locating uranium deposits. The radon content of some spa waters was often advertised as an indication of their curative properties. The wastes on the other hand are buried deep in the earth after being combined as a glass and sealed in metal containers. Rather than creating new radioactive hazards, the nuclear industry can be regarded as taking a natural radioactive material from the surface layers of the earth, extracting energy from it, and burying the waste deeper in the earth. This reduces not adds to the radioactive burden of man.

The over-concern with high level waste may have had the

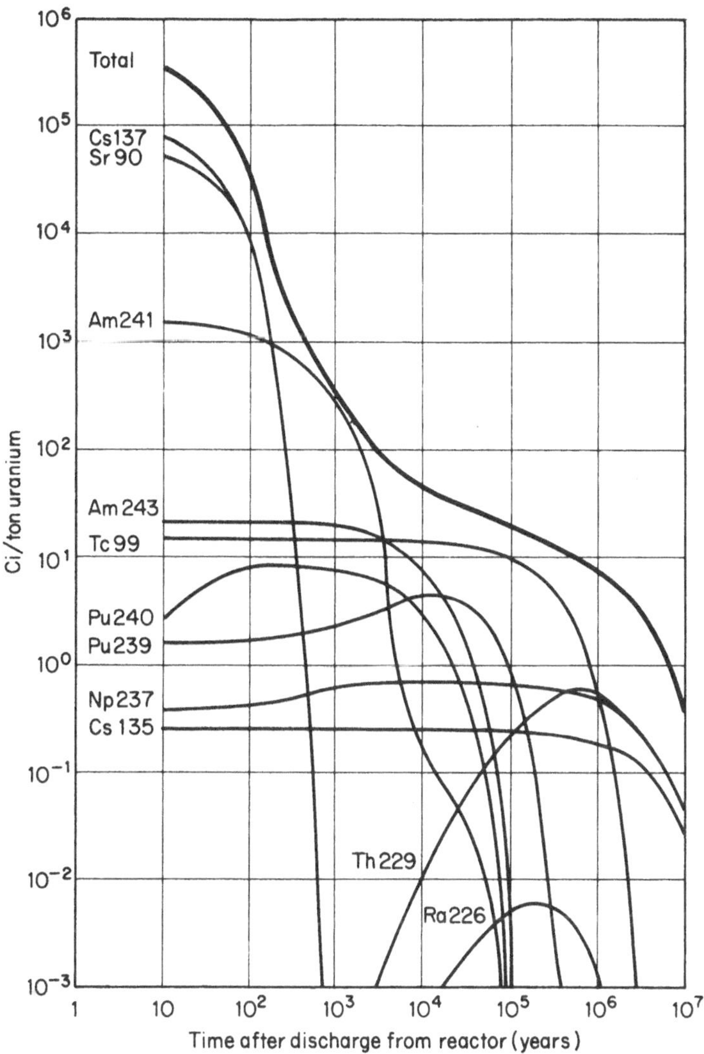

Figure 59: *Radioactive elements in high-level waste, assuming reprocessing takes place ten years after discharge of the spent fuel from the reactor. (Source: Ref. 2)*

disadvantage of diverting attention from the possibly more difficult but less dramatised problems of dealing with the much larger quantities of low level waste – more difficult only in that the volumes to be treated are much larger. These include ore tailings from uranium mining and milling, various reactor wastes mainly as ion-exchange resins and low and medium active

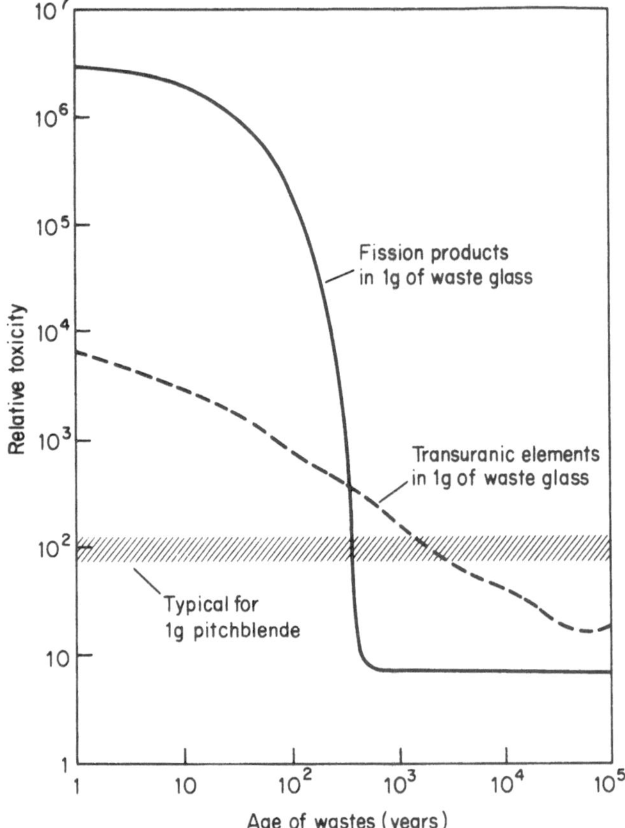

Figure 60: *Relative ingestion radiotoxicity of components of highly active waste and naturally occurring radioactive pitchblende. This assumes: PWR fuel, uranium fuel cycle, 1% plutonium in waste, 15% waste in solidified (glass) material, typical pitchblende toxicity normalised to unit weight. (Source: Ref. 3)*

reprocessing waste, including the zirconium fuel can hulls.

Reactor wastes, in quantity between some 200–500 cubic metres per year from a 1,000 MW reactor are mainly the ion-exchange resins from the clean up of the cooling system and fuel storage ponds together with laundry and some metal components. Most of this need only be kept for some decades before it can be handled as normal industrial waste. The more active material can be separated and treated by incineration to reduce the volume and fixing in bitumen or concrete, followed by land burial or sea disposal or as at the Asse mine in Germany by disposing underground in a disused salt mine.

The zirconium fuel cans which are chopped up when the fuel elements are dissassembled and leached in the reprocessing plant are contaminated with small amounts of undissolved fuel as well as activity arising from their irradiation in the reactor. The quantity could be between 150 to 350 kg per tonne of fuel processed in a volume of about 0.5 m^3. These present a particular problem in that zirconium in finely divided form can be pyrophoric. They are therefore normally stored under water in waste silos at the reprocessing plant for the short and medium term. For the longer term studies are being carried out to reduce the contamination of the hulls and their immobilisation in cement, glass or metal alloys or by conversion into silicates to give a material with very low leachability.

Even though they are naturally occurring materials, the tailings or wastes from mining and milling uranium containing minerals require special attention. In the extraction of uranium, the ore is crushed and ground and the uranium leached out by acid or alkali. The waste left is a slurry of finely ground solid, the mill tailings. This is pumped to a tailings pond where the tailings settle in a pile. The radioactive content of the tailings includes unextracted uranium and the uranium daughter products of which the most serious is radium-226. Although the radium was originally present in the ore, it has been brought to the surface and transformed into a finely ground material which significantly increases its potential biological hazard. This could come from the release of radon gas which will persist at a fairly constant level for thousands of years, and also from the inhalation of wind borne particles. To some extent an awareness of the danger gives some protection and these tailings should no longer be used for building purposes for which their finely ground nature had made them an attractive material. This is now the case in the US State of Colorado which prohibits the use of tailings as construction material.

One step that can be taken to reduce the radium hazard is to extract it from the tailings slurry by an additional extraction stage before discharge to the tailings pond. In some countries this is done on a routine basis. Another measure to minimise the release of radon and the airborne suspension of particles is to cover the settled tailings with a thick layer of earth. In essence this is equivalent to returning the radionuclides in the tailings below ground from which they came.

The costs of waste disposal would not be high relative to the value of the electricity produced, even with the elaborate procedures of storage, vitrification, curing and geologic disposal. They have been estimated by a working group of the International Fuel Cycle Evaluation study as about US$10–15 million per gigawatt year of electricity for the wastes from power plants with a capacity of 50 GW.[4] This can be compared with the value of the electricity, which at $0.1 per kWh would be US$876 million – about 1.5%. Although relatively small, these costs in absolute terms are a burden that society has to carry. It should be questioned how far they are justified from a realistic health and safety evaluation or whether they are unnecessarily high to meet the largely emotional need that absolute safety should be demonstrated. It may well be that the money could be better spent on other more urgent, more desirable social needs.

From the technical point of view, waste disposal methods require only the application or adaptation of existing technologies. Vitrification is already being applied on an industrial scale for low burn-up fuel and has been demonstrated for high burn-up fuel. The design of deep underground repositories has so far only been based on generic concepts and field experience. Some of the assumptions made would only be validated by the construction and operation of a repository. But it can be concluded that waste disposal can be carried out safely without undue risk to man or the environment.

The problems of waste disposal are then more political and administrative than technical. What is required is a clearly defined framework setting out the responsibilities of the different parties involved and the limits to which they are required to conform. The legal, administrative and financial aspects for the long term storage and disposal of waste need to be thoroughly examined. Some of the wastes and the spent fuel originate at the nuclear power station, other wastes arise when the spent fuel is reprocessed by a reprocessing company. In some countries waste handling organisations are being formed – The Nuclear Waste Disposal Bill (December 1977) put forward as a private members bill in the British Parliament proposed the formation of a Nuclear Waste Disposal Corporation. In many countries there is the assumption that the ultimate responsibility for the long term disposal of nuclear waste will pass to the State. Alternatively with proposals for regional fuel cycle centres which will undertake the reprocessing of nuclear fuel for a group of countries the solution may be found through international organisations. International agencies, regulatory authorities, government organisations, industry and the public must all be involved in arriving at a satisfactory solution.

REFERENCES

1. L.E.J. Roberts, Lecture to BNES, November 1978.
2. *Handling of Spent Nuclear Fuel and Final Storage of Vitrified High Level Reprocessing Waste*, KBS Report, 1978.
3. *Health Implications of Nuclear Power Production*, WHO, Copenhagen, 1977.
4. *International Nuclear Fuel Cycle Evaluation*, Report of Working Group 7, January 1980.

16
Nuclear safeguards and non-proliferation

" . . . in the long run there is no way of stopping the spread of nuclear technology amongst nations, and we must face the proliferation problems that result. The question is therefore not how to stop nuclear development but how best to make use of it and how to apply effective safeguards."[1]

Fissile material of the right quality can be used to cause a nuclear explosion; the first application of nuclear technology in the early 1940s was to produce atomic bombs; the first nuclear programmes in UK, USA, USSR and France were started for military purposes; a nuclear research programme could provide access to materials, equipment and technology that could be used to produce nuclear weapons. Yet there is no inevitable tie between a civil nuclear power station and a weapons programme. It is possible – and many countries have taken this step – to develop nuclear power for peaceful purposes and to renounce all intention to develop nuclear weapons.

There has so far been no case where a country has developed nuclear explosives by diverting material from a civil power station. In the 10-year period from 1945–1954 three countries (US, USSR and UK) developed nuclear explosives. Between 1955 and 1964 another two countries (France and China) and between 1965 and 1974 one single country (India) followed suit. Over this same time span the world nuclear generating capacity has grown from 5 MW in 1955 in one country, to 54,000 MW in 1974 in 19 countries. It is now 110,000 MW in 22 countries. From this alone it is evident that there is no relationship between the expansion of nuclear power and the development of nuclear explosives.

This point comes out even more clearly when it can be shown that for each of the five nuclear weapon states the explosion of their first bomb preceded, not followed, the entry into service of the first prototype commercial power reactors (Table 78).

The reasons for this are simply that it is much easier, cheaper and quicker to produce weapons grade fissile material in a special purpose built facility. All that is needed for a plutonium weapon is natural uranium, a research reactor, or other small reactor, a moderator which could be graphite or heavy water and a simple reprocessing plant. US studies have estimated the total cost at

Table 78: *Dates of first atomic bombs and reactors*

Country	Date first atomic bomb exploded in atmosphere	First prototype commercial power reactor (50 MW(e)) Date of service	Name
United States	1945	1957	Shippingport
USSR	1949	1958	Troitsk
United Kingdom	1952	1956	Calder Hall
France	1960	1964	Chinon
China	1964	No firm evidence of a power reactor in service	

about $50 million. The scale of effort is that of a large nuclear research project. On the other hand a nuclear power plant would require an investment of the order of $1,000–3,000 million, 20–60 times greater than the special weapons plant. It could only be built within a complex administration and regulatory structure. Large numbers of professional and specialist staff would have to be trained. For those countries lacking the necessary technological skills and industrial capacity the plant and its fuel would have to be imported and these would be placed under international safeguards agreements. The alternative of building an independent commercial enrichment plant to supply the fuel, and a commercial reprocessing plant to extract the plutonium would require a very much greater expenditure. If an enriched uranium weapon were to be developed a small plant specially built to produce highly enriched uranium would be less costly than a much larger plant to produce low enriched power reactor fuel at an economic cost.

The need to control nuclear weapons development was foreseen at a very early date. A first attempt was made in the United States Baruch Plan of 1946 which introduced the concept of physical international control combined with an international inspection, but this failed, mainly due to Russian opposition to the inspection measures, and by 1952 three nuclear weapon states, USSR, UK, USA, had exploded bombs. A fundamental change of strategy came with the unprecedented initiative of the Eisenhower "Atoms for Peace" programme of 1953 to share the benefits of the peaceful application of nuclear technology in return for the acceptance of agreements and controls to prevent military use. This programme led to a large scale, world wide transfer of peaceful nuclear technology. Many thousands of scientists from the developing countries were trained in the US and other industrialised countries. Nuclear research centres were established in many developing countries and these stimulated a marked growth in the application of science and technology in all fields. The countries that benefited most from this programme were India, Argentina, Brazil, South Korea, Taiwan and Pakistan. India had set up its own Atomic Energy Commission as early as 1948. In the course of implementing this policy the International Atomic Energy Agency was created in 1957 to promote and to control the peaceful use of nuclear energy.

Independently of these peaceful initiatives "horizontal proliferation" – the spread of nuclear weapons capability – occurred when France in 1960 and China in 1964 exploded their first atomic bombs. One effect of the Chinese test was to bring about a change in Soviet policy which had hitherto opposed attempts at weapons control; it now found a common interest with the US and other countries in trying to restrict further horizontal proliferation. This united front within the framework of the Eighteen Nation Disarmament Committee 1962–1968 led to the formulation of the Treaty on the Non-proliferation of Nuclear Weapons (NPT), which was opened for signature in 1968 and came into force in 1970.

NPT required that the entire nuclear effort of states party to the Treaty should be covered by safeguards. The Non-Nuclear Weapon States (NNWS) which adhered to the Treaty committed themselves not to acquire nuclear explosive capacity. In return for renouncing nuclear weapons and for accepting the application of safeguards across their entire national nuclear industry the NNWSs were offered commitments under Article III.3 and Article IV that their peaceful nuclear development would not be impeded, and that they would have the right to take part in "the fullest possible exchange of equipment, materials and scientific and technical information on the peaceful uses of nuclear energy". Those other parties to the Treaty in a position to do so accepted a commitment to help their less advanced companions to develop their peaceful nuclear energy programmes.

The Non-Nuclear Weapon States, although many of them have subsequently adhered to NPT, always believed that they had the worst of the bargain, in that the concessions and restrictions were all on their side. They would argue that "vertical proliferation", the increasing stockpiles of nuclear warheads of ever-increasing sophistication, is the greater danger to world peace. In August 1968 the NNWSs took the initiative of convening a conference in Geneva under the auspices of the UN, and numerous resolutions were passed calling for overall disarmament measures such as the creation of nuclear-weapon free zones, cessation of production of fissile materials for weapons purposes, the reduction and eventual elimination of all stockpiles of nuclear weapons. All these failed to make any impact on the Nuclear Weapon States and the NPT text was not modified in any way. Even the limited goal set in Article VI:

> "Each of the Parties to the Treaty undertakes to pursue renegotiations in good faith on effective measures relating to the cessation of the nuclear arms race at an early date and to nuclear disarmament, and on a Treaty on general and complete disarmament under strict and effective international control."

is as far off today as it was twelve years ago.

Despite these limitations NPT has attracted the adherence of 110 states. These include three weapon states (UK, USA, USSR); France, for some singularly French reason, has stated that although it will not become a party it will pursue policies that will in all respects be consistent with the Treaty, and it has done so. The 5th NWS, China, has expressed opposition to the Treaty but does not appear to have taken any actions that are strikingly inconsistent with it.

The calm which NPT had engendered was shattered in May 1974 by the Indian test of what was claimed as a peaceful nuclear explosion. Many myths and half truths have been attached to this event that should be dispelled. In the first place the Indian explosion was not, as many have tried to suggest, a failure of NPT. India was not a party to the Treaty. The test was carried out by the use of nuclear facilities and material that had never been placed under safeguards. Secondly at that time – the early 1970s – there was a considerable interest in the peaceful use of nuclear explosives, as typified by the US "Ploughshare" programme for seismological, oil and gas stimulation, rock breaking, constructional and other purposes. The US programme was abandoned in 1973 but the USSR has continued with between 2--5 tests a year for these "peaceful" purposes. And as has been said in a recent review "India is probably one country that could profit considerably from nuclear explosives engineering".[2] It was then not unreasonable at that time for India to carry out such a test, said to be for the excavation of underground water storage reservoirs. A third point is the implication often made that this was a shock event which took the rest of the world by surprise. In fact the Indians did not conceal their intentions and an official government statement that such a test was being prepared was made in 1971 almost three years before the test took place.

If there was any failure, it was on the part of the major countries to apply any significant diplomatic pressure to persuade the Indians that it would not be in their best interest to carry out such a test at that time.

The SIPRI Year Book (Stockholm International Peace Research Institute) for 1978 gave an account of the events which are presumed to have led to South Africa abandoning a supposed nuclear test in August 1977. By plotting the detailed tracks of Soviet Cosmos satellites 922 and 932 and of the American Big Bird reconnaissance satellite SIPRI shows how they passed repeatedly over the suspected nuclear test site in the Kalahari desert. According to this account the USSR informed the US that South Africa was secretly preparing to detonate a nuclear explosion. France, Federal Republic of Germany and the UK were also informed. After considerable diplomatic activity it was announced that South Africa had promised that "no nuclear explosive test will be undertaken . . . now or in the future". In their analysis of these events SIPRI concluded that the fissile material for the South African weapons must be highly enriched uranium from their own production; it could not be plutonium from the SAFARI-1 research reactor, since the fuel from this is sent to the US or UK for reprocessing under International Atomic Energy Agency supervision. There can then be little doubt that strong, combined diplomatic activity by the major powers could have prevented the Indian explosion.

One effect of the Indian explosion however was that it led to a revival of the earlier policies of restriction and diverted non-proliferation policies away from the fruitful collaboration of the Eisenhowever "Atoms for Peace" programmes. This upset the essential bargain of NPT, whereby in return for renouncing all intention to manufacture nuclear weapons, and accepting the IAEA safeguards system, the Non-Nuclear Weapon States were offered full assistance under Article IV.

These restrictive policies were formulated in the rules of what has become known as the London Suppliers Club, and also in the legislation of the US Non-proliferation Act of 1978. The London Suppliers Club was formed in 1975 with an initial membership of UK, France, West Germany, US, USSR, Canada and Japan: this was later widened with the addition of Netherlands, Sweden, Switzerland, GDR, Poland, Czechoslovakia, Italy and Belgium.

Within this wider membership there are differences of approach since only four members (US, France, Canada and USSR) can claim to be wholly self-sufficient in nuclear technology and materials, the others are to a greater or lesser degree both suppliers and recipients. The approach of the London Club, as set out in its guidelines published in 1978, is based upon restricting transfer of basic information and know-how for the "sensitive" technologies of enrichment, reprocessing, fuel fabrication and heavy water production. The guidelines further stipulate that supplied nuclear material and equipment must be subject to IAEA safeguards and inspections and that no material or equipment should be transferred to a third country without the concurrence of the supplier nation. This is perceived by the recipient countries as a policy of denial. But the history of nuclear energy has repeatedly shown that a policy of denial can at best only provide a temporary delay, and often serves to stimulate the development of the restricted technology. For instance, in 1944 the US denial of access to plutonium technology to the UK and Canada led these two countries to develop a new reprocessing method which is the basis of the now widely used Purex process. Although enrichment processes are still classified as secret this has not hampered the spread of enrichment technology with the independent development in a number of countries of a variety of different methods.

Sales or transfers that are dependent on technology acquired from one of the major suppliers would fall within the scope of the London Club. But it is difficult to establish any hard dividing line beyond which the input of national technology can be regarded as the major factor. In the civil nuclear industry for instance, KWU which started with a Westinghouse PWR licence, has evolved what is recognised as an independent German design of PWR.

Nuclear knowledge and know-how is already widely spread and a third tier of nuclear suppliers is now emerging, from among the advanced developing countries. These include Argentina which has provided a large research reactor to Peru and technical assistance to Ecuador, Paraguay, Chile and Colombia; India which has trained Egyptian, Iranian and Vietnamese nuclear scientists; Brazil, which is discussing the supply of nuclear technology to Iraq in return for much needed oil, and India similarly with Libya. The capabilities of these countries will grow. India which is now building its own nuclear power stations could offer them for export. Brazil is now building up a complete nuclear manufacturing capability through a number of joint Brazilian–German companies that have been established in Brazil under the nuclear co-operation agreement with Germany. Korea, although it does not have one single overall partner, is following a similar route to Brazil in establishing as quickly as possible a nuclear design and manufacturing capability. This is being done by a number of joint venture and licence agreements which include the training of Korean staff as well as the supply of equipment. There are also South Africa, which can supply natural and enriched uranium, and Spain, with a complete capacity to produce heavy reactor components and to manufacture fuel. India, and (according to the 1978 SIPRI year book) Israel have small plants for fuel reprocessing and Argentina is now constructing one. These countries could conceivably become exporters of fuel reprocessing technology in the future. India is also developing fast breeder technology.

It is however not only the London Suppliers Club which is trying to impose a policy of denial. This is also the main principle of the US Nuclear Non-Proliferation Act. President Carter made the point:

> "If we felt that the provision of atomic fuel was being delivered to a nation that did not share with us our commitment to non-proliferation, we would not supply that fuel."[3]

The Act makes compliance with the US non-proliferation requirements a necessary condition for the supply of low-enriched uranium. It is also clear that the intent of the Act is to restrict the spread of reprocessing and to discourage commitment to plutonium fuels for recycle or breeders: it goes beyond the usual safeguards criteria and imposes new conditions on the US trading partners, in many cases doing so retroactively even where exports were covered by existing contracts and agreements. While such unilateral action by the US is pointed to by the Third World countries as evidence of the unreliability of the major supply countries, the Act seems to be aimed also at those countries, France, UK and Japan as well as India that are actively seeking to develop the fast reactor (as are the USSR and its Comecon partners). But the ability of the United States to enforce its policy by withholding supplies of enriched uranium now seems doubtful.

A considerable over-supply of enrichment arising from the dual effect of the entry of new suppliers into the market and an unexpectedly high rate of slippage of planned nuclear programmes has led to a lessening of dependence upon US suppliers.

It might then be thought that the "supplier leverage" might pass to the largest potential suppliers of natural uranium, Canada and Australia.

But there are many other suppliers, notably in Central and South Africa, and with greater exploration new sources will no doubt be brought in. In any case any threatened shortage of uranium would have the immediate effect of accelerating breeder reactors and the recycling of plutonium in thermal reactors, the two technologies which are most suspect under the present US proliferation attitudes.

One particular objection to a policy of denial is its essential discriminatory nature between the haves and the have-nots. This is particularly felt by the developing countries who

> "view this step as yet another example of an organised and deliberate effort on the part of the advanced countries to deny technologies to the developing countries and impose onerous restrictions on its transfer which is tantamount to constricting the growth of their civil nuclear programmes".[4]

The distinction between weapon states and non-weapon states is becoming progressively less important; the real difference now lies between supplier and customer countries, between the haves and the have-nots of nuclear technology. This view is forcibly expressed in the final communique from the Havana Conference of Non-Aligned States in September 1979:

> "The Conference affirmed the inalienable right of all States to apply and develop their programmes for the peaceful uses of nuclear energy for economic and social development in conformity with their priorities, interests and needs. It deplored the pressures and threats against developing countries aimed at preventing them from pursuing their programmes for the development of nuclear energy for peaceful purposes.
>
> All states should have unhindered access to and be free to acquire technology equipment and materials on a non-discriminatory basis for peaceful uses of nuclear energy, taking into account the particular needs of the developing countries.
>
> The Conference stressed the need for observance of principles of non-discrimination and free access to nuclear technology and reaffirmed the right of each country to develop programmes for the use of nuclear energy for peaceful purposes in conformity with their own freely determined priorities and needs.
>
> Concern for non-proliferation should not be used as a pretext to prevent States from exercising the right to acquire and develop peaceful nuclear technology. The Conference expressed its concern at the monopolistic policies of nuclear supplier countries restricting and limiting the transfer of technology and imposing conditions which are incompatible with the sovereignty and independence of the developing countries. It called for full observance of the principles mentioned above for the utilisation of nuclear technology for peaceful purposes, which have been endorsed unanimously by the United Nations General Assembly."

Some of these developing countries, which are now emerging as new industrial powers, have rapidly growing energy demands. They are conscious of the fact that they have to build up their industrial structure at a time when energy is expensive: whereas the present industrial states had the good fortune to develop their industries on cheap energy. They need large quantitiesof the cheapest power they can obtain and this is often nuclear power. Table 41 shows that in 1979 some of the developing countries were already taking a greater share of their electricity from nuclear power than some of the industrial countries. These countries are then resentful of the fact that their need for nuclear power does not seem to be fully appreciated. Although many of them are following with keen interest the results of research and development work on the alternative energy sources – solar, wind, biomass etc. – which could be utilised in remote and rural areas they are allergic to any attempts to have the "soft" energy technologies pushed onto them.

> "They realise that these new technologies are not yet sufficiently well developed to provide solutions to their near term energy problems, particularly for meeting large scale electric power needs for industrial purposes . . . and regard such moves as yet another way of persuading them to give up their interest in nuclear power programmes."[4]

This suspicion that for the developing countries nuclear power leads to nuclear proliferation is reflected in the US Non-Proliferation Act of 1978. Section 2(d) of the Act affirms that it is the policy of the United States "to identify alternative options to nuclear power in aiding (other) nations to meet their energy needs". This is intended to encourage developing countries to abjure nuclear power and to turn instead to sources such as solar power. However, the pursuit of solar power or other "alternative options" does not in any way bar the road to nuclear explosives. Israel is probably the country most advanced in the world in the actual use of solar power: Israel has no nuclear power plant in operation or even under construction: yet, it is generally accepted that Israel has the "direct capability" to manufacture nuclear explosive devices.

There is then a growing consensus of informed opinion that non-proliferation policies cannot be enforced by a policy of denial which might indeed have the opposite effect and stimulate the creation of independent capability. Restrictive measures could only increase the existing feelings of mistrust between the advanced and the "Third World" countries. This is indeed one of the general conclusions that has emerged from the INFCE study – a massive 2 year effort in which 66 countries and 5 international organisations took part in one of the most comprehensive evaluations of a technology ever undertaken. INFCE accepted that it would be desirable for reprocessing to be carried out in a relatively small number of large plants, and that fast reactors were likely to be of interest mainly to countries with large nuclear power programmes. But any suggestion that there are 'big' nuclear countries which should, and 'little' nuclear countries that should not have access to the benefits of more advanced (and more sensitive) technology would only arouse the resentment and hostility of the smaller industrial countries and the developing world. INFCE also recognised that proliferation is basically a political matter; while there is no fuel cycle that is entirely proliferation resistant, there is at the same time no fuel cycle that cannot be reconciled with a non-proliferation regime given improved institutional arrangements and improved safeguards.

What is required is a set of non-discriminatory rules, under which nuclear commerce and transfers of technology can proceed, which would restore confidence in the basic principle of NPT whereby the exporting countries will give irrevocable guarantees of supply in return for the acceptance by the importing countries of a strict non-proliferation policy coupled with comprehensive and irrevocable safeguards measures. This could best be achieved by building upon the present NPT Treaty and supporting it with stronger political and diplomatic measures to ensure that proliferation is contained, by paying special attention to the problems and needs of those countries who are still outside the NPT. The extent and nature of the problem can be seen by considering the countries with some significant nuclear activity that are as yet outside the Treaty. They consist of India, Pakistan, Israel, Egypt, Indonesia, Turkey, the Democratic Peoples Republic of Korea, China, Argentina, Brazil, Chile and Colombia, South Africa and Spain. Of these the Latin American countries could be covered by the expected ratification of the Tlatelolco Treaty which aims at creating a nuclear weapons free zone in Latin America. Spain may be obliged to accept the NPT when it joins the European Community since it would then in any case come under the Euratom safeguards agreement. South Africa, as has been seen, can be persuaded not to demonstrate nuclear weapons capability. There is no obvious reason why Turkey and Indonesia should abstain, and if, as has been reported, Turkey is to import a nuclear power station from Sweden and/or Russia it will have a stronger incentive to join NPT.

This leaves two sensitive areas of political tension. The Middle East and the Far East. In the Middle East the rapprochement between Israel and Egypt should reduce the tension and could pave the way to a wider settlement. In the Far East much will depend upon the role of China. If the post-Mao emergent China will give a positive lead in reducing tension this could greatly reduce the spread of nuclear weapons. If India were assured of a peaceful Chinese neighbour, relations with Pakistan could in turn be improved. In a similar way Chinese policies could influence North Korea and indirectly South Korea and Formosa.

In addition to the IAEA safeguards and NPT, together with continuing diplomatic and political efforts to reduce international tension there is now a move to create a third barrier to the spread of nuclear weapons with proposals to form international or regional groupings to operate or control some of the more "sensitive" aspects of nuclear technology.

A first step in this direction is now being taken with discussions being carried out by the IAEA and a number of its member states on the implementation of an international plutonium storage scheme. Such a scheme for the storage, and subsequent release for approved purposes of excess plutonium, was foreseen in the IAEA Statutes (Article XII A 5). Its aim is the physical international control of plutonium at the most sensitive stage of the fuel cycle – after reprocessing and before use.* If, as is expected, the international plutonium storage can be successfully implemented it would pave the way towards further steps to cover other stages of the fuel cycle, such as fuel fabrication where the plutonium is still relatively accessible, and reprocessing where the availability of regional or international plant, without discrimination to the users, would go some way to discourage countries from setting up their own national facilities at a higher cost for a small throughput. In this way, combined with safeguards on reactors – where in any case the fuel is inaccessible because of the high radiation levels – the reassurance provided, that fissile material is not being diverted to military use, can be brought to a high degree. But it must be emphasised that there is no single, magic instant solution to the problem of nuclear weapons proliferation. INFCE has shown there is no technical "fix" – no proliferation resistant fuel cycle; experience tells that there is no political "fix". The approach must continue to be a range of interlocking, related measures supporting one another. Each such measure should be practical and positive in itself; but in the final analysis it is the total non-proliferation regime which counts.

REFERENCES

1. S. Eklund, Statement to 21st General Conference of the IAEA, September 1977.
2. D. Davies, *Peaceful Nuclear Explosions: Nuclear Energy and Nuclear Weapon Proliferation*, SIPRI, 1979, p. 300.
3. White House Briefing Room Transcript, April 1977.
4. Munir Khan, Nuclear Energy and International Co-operation: A "Third World" Perception of the Erosion of Confidence, The Rockefeller Foundation/Royal Institute International Affairs, September 1979.

* International Plutonium Storage was one of the institutional arrangements which was strongly supported by INFCE.

17
Opposition to nuclear power

Despite the promise of nuclear power in providing a clean, safe and reliable source of power for the foreseeable future there are in many countries delays and hesitations in accepting this new source of energy. These are due in a large measure to the hostility of part of the public to any nuclear developments.

This opposition first manifested itself in the USA at the beginning of the 1960s. Although the early power stations, Dresden in service 1960, Yankee Rowe 1960, Indian Point 1962 and Humbolt Bay 1963, were built and came into operation without arousing any unease, the seeds of mistrust were first sown when the US Atomic Energy Commission over-ruled their own Advisory Committee on Reactor Safety and authorised the construction of the experimental Fermi fast reactor near Detroit. This "imprudent and arrogant" action led to the opening in January 1957 of the first full scale inquiry into a nuclear project which ran for some eight months. Another source of future trouble was the publication by the US AEC in March 1957 of the study it had commissioned on the risks and consequences of nuclear accidents, the now notorious Brookhaven report (Wash-740) which caused alarm by suggesting high mortality and damage figures, but at a very low probability: this was eagerly seized upon by the opponents of nuclear power who dramatised the consequences whilst ignoring the low probabilities.

The first signs of local opposition to the building of nuclear stations came in 1961 with the Pacific Gas and Electricity Co. plans for Bodega Head on the coast north of San Francisco. This site was not only within a local nature reserve but also on a main earthquake fault. PG & E were embroiled again in a similar controversy in 1964 over the Malibu Beach site for exactly the same reasons. These events motivated the influential ecological group the "Sierra Club" to come out in strong opposition to nuclear power. The dilemma facing the utilities was that remote sites with adequate cooling water were most readily found in the dwindling number of unspoilt recreational areas, while the alternative of urban siting was also unacceptable to the public.

Fears of nuclear power and of the cancer risk and genetic

danger of radiation were aroused and intensified by the growing concern over fallout from atmospheric tests of nuclear weapons during the second half of the 1950s which eventually led to the atmospheric test ban of 1962. By this time the literature of the growing ecological movement against pesticides and other pollutants typified by Rachel Carson's *Silent Spring* found echoes in anti-nuclear power books such as *Perils of the peaceful atom – the myth of safe nuclear power plants* by Elizabeth Hogan and Richard Curtis 1969 and *The Careless Atom* by Sheldon Novick in 1970. A new twist was given by Professor Sternglass, a health physics professor at the University of Pittsburgh with a supposedly scientific article published in *Esquire* in 1969 with the emotional title "The Death of All Children". This proclaimed the so-called Sternglass effect which asserted that a slowing in the rate of decrease of still births in the US could be correlated with weapons test fallout, and later in 1971, he extended this to cover similar supposed effects in the neighbourhood of operating nuclear power stations. This work has been totally rejected by the scientific community as the "arbitrary selection of data supporting his hypothesis and ignoring of those that do not".[1] The mounting sensitivity over the need to protect the environment led to the US National Environmental Policy Act of 1970. And in the Calvert Cliffs decision of 1971 the courts ruled that the AEC had violated the Act by not taking into account the non-nuclear effects of thermal discharge when authorising the construction of the plant and revoked the authorisation. This led to an expensive delay before the plant could eventually be completed.

The long drawn out public hearings which were then invoked for all nuclear authorisations not only caused delay with consequent increase in costs, but also provided a public platform for the opponents to put forward their views, to cross-question the nuclear scientists and technologists and to uncover real or imaginary problems which were then dramatised to generate widespread fears on the use of nuclear technology.

In addition to established environmental groups which increasingly adopted anti-nuclear policies a large number of organisations were formed on both the national and local level specifically to oppose the construction of nuclear power stations. In some cases organisations formed for one purpose, such as the anti-Vietnam war women's protest group "Another Mother for Peace" shifted their focus to the anti-nuclear movement when the end of the Vietnam war ended their original purpose. The whole US anti-nuclear movement, consisting of groups differing widely in size, objectives, organisation and purpose was then brought together with the formation in 1973 by Ralph Nader of an umbrella organisation, Critical Mass, to lobby politicians in Washington and the State Capitals, to seek television time for anti-nuclear pronouncements, to work with churches and other special interest groups and to intervene in the courts on a wide array of nuclear issues. The US administrative procedures lend themselves readily to legal action: the National Resources Defense Council which has on its staff 14 attorneys and 4 scientists is particularly active in filing suits on nuclear issues. In this way an anti-nuclear movement has emerged; well organised, well financed and with full time staff members. With a centrally directed policy it has been able to establish small activist groups within a number of existing reputable organisations which are then persuaded to adopt anti-nuclear resolutions, so that the anti-nuclear movement is strengthened by the apparent support of independent, unbiased and often influential organisations. A well known example is the National Council of Churches resolution opposing the "plutonium economy". Similar resolutions have been put forward within the League of Women Voters, the American Medical Association and the American Association of University Women.

This course of events in the USA has, with a delay of 4–5 years been more or less closely followed in most of the industrial countries.

A novel European feature has been the introduction of the occupation of nuclear sites as at Kaiseraugst in Switzerland, sometimes leading to violence, particularly in Germany, where pitched battles took place between demonstrators and police at Whyl and Brokdorf, and at Creys Malville in France one of the demonstrators died.

It could be significant that in the poorer, energy hungry world, India, Pakistan, Korea, Brazil, Argentina etc. there is no widespread protest despite the size of the nuclear programmes which rival or even exceed those of some of the industrialised countries.

It is apparent that there are close, even if informal links, between the anti-nuclear groups in all countries. Leading figures from the US protest groups have been prominent in Europe, particularly in Scandinavia and in the UK at the Windscale Inquiry. UK representatives have taken an active part in anti-nuclear meetings in Australia. Germany – Austria – Switzerland have always had a close network and this is being extended to include the Netherlands, Scandinavia, France, Italy, Spain and UK. This network has evolved an efficient means of communication and a free flow of technical information. It has been claimed that this has been facilitated by the involvement of the institutions of science and of the church.[2]

Yet despite the apparently widespread support it seems that the core of anti-nuclear movement is made up of a very small number of individuals. It has been established that the Union of Concerned Scientists which was on its own largely responsible for generating alarm over the Emergency Core Cooling System and other safety issues, which may have deflected the industry and the regulatory authorities from a more rational approach to safety considerations,* "boils down to 12–20 people".[3] Another US group, Environmental Action, which claimed the support of 8,000 individuals had its policy determined by 8 or 10 paid staff members. The Citizens Association for Safe Energy with a mailing list of 2,500 supporters had a resolution calling for a complete halt to all US nuclear power plants drafted and adopted by a 3-man board of directors. Similarly in the UK the lead anti-nuclear organisation, The Friends of the Earth Ltd, which is an associate of the US body, has admitted that its anti-nuclear policy was determined by a few key individuals, and then promoted throughout its 5,000 members by circulating policy documents and by discussion seminars.[4] As already noted the activists can then penetrate other organisations so that their views can be given the tacit support of a much wider membership. It was remarkable that the Durham County Council, The Town and Country Planning Association, the National Council for Civil Liberties and the British Council of Churches not only took an active part in opposing the Windscale reprocessing plant at the Public Inquiry held in 1978 but some even went to the expense of being represented by Counsel and bringing in witnesses from the USA to support their case. Yet these are bodies whose responsibilities would normally appear to be quite remote from Windscale, from matters of energy policy and from the wider issues of public health and safety.

It then seems that the opposition to nuclear power manifests itself at two different levels. At the public level there is a general pool of unease associated with a wide range of conscious and unconscious fears — fears of radiation, of cancer, of genetic damage, of the atom bomb, of big business and big government and of a too rapid rate of change of society leading to a mistrust

***The Presidents Commission on the TMI Accident has criticised the NRC for having focussed most attention on large-break loss of coolant accidents rather than on the smaller, more probable accidents such as TMI. But in doing this the NRC were only bowing to the public clamour raised by the UCS on major break accidents.**

of the applications of science and technology. The role of fear as an important factor in society has been analysed and commented upon by many psychologists, philosophers and theologians.* The receptivity of the public to these fears is then being deliberately, perhaps even cynically, manipulated by a much smaller group of people who, dissatisfied with the present industrial society are seeking to change what they regard as its materialistic acquisitive values and to impose an "alternative" society.

The probability of an adverse public reaction against the introduction of nuclear power was first considered in a perceptive study by the World Health Organization whose report *Mental Health, Aspects of the Peaceful Uses of Atomic Energy* was published as long ago as 1958 and has since been ignored or forgotten[5]

> "it is logical to infer that mankind's encounter with a source of energy of such shattering possibilities as atomic power will cause strong psychological reactions and that some of these will probably have to be considered more or less pathological"

and it noted that in particular

> "Radiation energy is indeed liable to arouse anxious associations on account of man's first experiences of contact with it. In medicine, ionising radiations were first used for the diagnosis and treatment of some of the most awe inspiring disease, like tuberculosis and cancer. On the other hand, humanity's first contact with nuclear power was the explosion of the first atom bombs. There is indeed evidence that it is exceedingly difficult for many people to keep the productive uses of atomic energy clearly separated in their minds from its destructive possibilities and that this inability contributes to making the whole concept of atomic energy potentially a frightening one."[5]

The WHO report also considers the effects of changes in society as in the industrialisation of Europe and North America over the past 150 years when the change from agriculture and commerce to a predominantly industrial society brought about a series of interconnected changes of far reaching effect. These effects are now becoming more pronounced with the accelerating applications of technology which are well described in books such as Alvin Toeffler's *Future Shock* and whose resultsarc so strongly deplored by Rozack *Where the Wasteland Ends* and Jacques Ellul in many writings, for instance, *The Technological Society*, first published in France in 1954.

This latent hostility to all new technology seems to be a fundamental social or psychological attitude. It is not new. It contributed to the Luddite movement, and also appeared at the time of

* Particular reference is made to the book by Jean Delumeau, *La Peur en Occident*, XIVe–XVIIIe siècles (Fayard, Paris).

the introduction of the railways which were, in the 1820s, opposed in terms as violent and irrational as are now used against nuclear power. These were described in a history of the railways published in 1851:[6]

> "Every report which could promote a prejudice, every rumour which could affect a principle, was spread. The country gentleman was told that the smoke would kill the birds as they passed over the locomotive. The public were informed that the weight of the engine would prevent its moving; and the manufacturer was told that the sparks from its chimney would burn his goods. The passenger was frightened by the assertion that life and limb would be endangered. Elderly gentlemen were tortured with the notion that they would be run over. Ladies were alarmed at the thought that their horses would take fright. Foxes and pheasants were to cease in the neighbourhood of a railway. The race of horses was to be extinguished. Farmers were possessed with the idea that oats and hay would no more be marketable produce; cattle would start and throw their riders, cows even, it was said, would cease to yield milk in the neighbourhood of one of these infernal machines."

This now seems quaintly amusing, but at the time it was taken seriously and public opposition combined with sectional interests to obstruct completely or delay the construction of some railway projects by a number of years. The Parliamentary Acts and special regulations under which the railways in the UK still operate are a legacy from this controversy.

With the belief that art gives valuable insights into history, attention is drawn to George Eliot's novel *Middlemarch*, first published in 1872, which weaves into fiction an account of the anti-railway movement, together with the attitudes and prejudices of the time. "In the absence of any precise idea as to what railways were, public opinion . . . was against them." The anti-railway activists are described as "stimulating suspicions" in the minds of the local populace. And with a striking reference, as apt today as it was then, to the readiness with which these suspicions were accepted – "Nettle seed needs no digging".[7]

As the pace of technological change has accelerated over the last 100 years there are now many in whom a call for a slowing down and turning to other values strikes a responsive chord. A whole school of anti-technological philosophy has developed under the leadership of Herbert Marcuse, in which technology (or science) is personified as a Force in its own right and equated with the totalitarian society:

> "Technology serves to institute new, more effective, and more pleasant social control and social cohesion. The totalitarian tendency of these controls seems to assert itself in still another sense – by spreading to the less developed and even to the pre-industrial areas of the world, and by creating similarities in the development of capitalism and communism.

> In the face of the totalitarian features of this society, the traditional notion of the 'neutrality' of technology can no longer be maintained. Technology as such cannot be isolated from the use to which it is put; the technological society is a system of domination which operates already in the concept and construction of techniques."[8]

The full meaning of totalitarian is expanded in a later passage . . .

> "By virtue of the way it has organised its technological base, contemporary industrial society tends to be totalitarian. For 'totalitarian' is not only a terroristic political coordination of society, but also a non-terroristic economic-technical coordination which operates through the manipulation of needs by vested interests."[8]

This fear of new technology, as might be expected, appears to be growing with the increasing pace of technological development. It is reflected in that not insignificant segment of our culture which tries to look into future - science fiction. A recent review of science fiction under the heading *Dark Trends – Psychology, Science Fiction, and the ominous consequences*[9] concludes that many science fiction authors predict that the world will continue moving towards an environment characterised by a combination of high population density and advanced technology; it draws attention to the agreement between artistic vision and scientific research – that the conditions of human civilisation are becoming more unpleasant and deleterious.

It seems probable as the WHO study suggests, that societies have a certain threshold of tolerance for rate of change, which if exceeded leads in some measure to social disorganisation. This could possibly be one of the underlying factors of the revolution in Iran, where the Shah conscious of the limited life of the oil reserves, the main source of the country's income, had been trying to transform the country into an industrial state at a rate too rapid for the mass of the people to accept.

The social disintegration is not so much an end state but a process of change which leads to an increase in irrational emotional states and unsatisfactory human relationships.

> "One's attention is attracted by the appearance at times of a kind of self-perpetuating downward spiral in which the psychological states generated by the disorganisation render the people less and less able to utilise whatever assets are available to them and to find some satisfactory form of solution. Sometimes whole societies are affected in this way; sometimes it is a question of sub-groups within societies . . . "[5]

The rejection of nuclear power at a time of increasing energy shortage would appear to be a good example of this process.

A similar reaction occurred at the time of an earlier energy crisis, in England during the 16th century, when a growing shortage of firewood led to severe hardship. The price of firewood which had remained stable until the 1540s increased by a factor of 4 by 1580 and 10 by 1620. Heavy penalties were then imposed for the unlawful cutting of wood. Those that suffered most were first the poor and second the small but growing industry, using wood fuel on an increasing scale, whose output was restricted by law. Yet at that time there was a hostile public reaction to the introduction of coal as a domestic and industrial fuel, with laws being enacted to prohibit its use. There were complaints about the smoke and soot and "nice dames" refused to enter a room heated by coal or to eat food cooked by it. Those who promote alarmist fears about civil liberties being eroded with the use of nuclear power should note that the first casualty of a severe fuel shortage is the freedom of the individual, whose right of access to fuel supplies is limited by law.

There are a number of instances of periods of social disintegration which have occurred in different societies in the past, while for the present times there is the comment by George Steiner in the Bronowski Memorial Lecture 1978 that there are: "three times as many registered astrologers in Western Europe and the United States as there are members of professional associations of physics and chemistry". It is suggested that at times of crisis the human mind becomes more susceptible to myths: "one of the realities of the world as tangible and real as material facts . . . Many scientists deceive themselves on the irrational nature of myths, and believe that it is only necessary to expose them to the light of truth and they will dissolve as a mist is cleared by the rays of the sun. But psychological analysis has shown that the myth is too firmly embedded in the human spirit for it to be dispersed so easily."[10]

The ready acceptability of the irrational, particularly in relation to the fear of death – notably by cancer – which, it is claimed,[11] has almost reached a hysterical level in some countries where it is said that cancer is a consequence of industrialisation and urban civilisation. Yet epidemiological studies by the International Cancer Research Centre suggest that for subjects of the same age the frequency of cancer is no greater in the industrial countries than in the developing countries. This whole subject of the refusal to accept facts, of a wish to embrace the irrational has been studied at length by M. Tubiana in his important book *Le refus du réel* (Editions R. Laffont 1977). As Marcel Proust wrote some 50 years ago "Les faits ne pénètrent pas dans le monde où vivent nos croyances, ils n'ont pas fait naitre celles ci, ils ne les détruisent pas".

These are factors which could explain the readiness of part of the public to respond to fears about the use of nuclear power which are being assiduously promoted by a small group of people who form the so-called environmental movement.

There has been some speculation on the motives of the "environmentalists" and their source of finance. Professor Hoyle[12] has suggested that it is part of a Communist inspired plot to attack the West at its most vulnerable point – energy supply. This seems unlikely if only because the Russians and the other CMEA countries now have one of the largest and most rapidly increasing nuclear power programmes in the world and would not wish to see it disrupted by anti-nuclear movements spreading from the west. The Russians for their part see it differently:

> "There are no scientific reasons to support the idea that nuclear plants are dangerous. Scaremongers in the West have completely different and highly political reasons. The opponents of disarmament are doing everything in their power to distract public opinions from nuclear weapons and concentrate it on nuclear power."[13]

It has been rumoured that rival energy industries, coal and oil have been financing the nuclear opposition, but again this seems unlikely since the major oil companies in their role of energy suppliers are expanding into both the coal and nuclear industries.

The roots of the environmental movement seem to lie more with those who are disaffected with, and seeking to change, the present industrial society.

The industrial revolution is regarded as an aberration from which society should be rescued. The arguments of the early guild socialists of the first decades of this century are rehearsed once more. The romantic poets are their guides. Bigness must be opposed; small is beautiful. The ideal society is to be formed of small self-governing communities living in harmony with nature and using[14]

> " . . . the relatively simple technologies that rely on natural energy flows, sun, wind, vegetable that are matched in scale and energy quality to end-use needs."

Technology will only be allowed if it is "appropriate" technology used to the full for the good of mankind. "Appropriate" is not defined, but it is the opposite of inappropriate technology:[15]

> " . . . inappropriate technology includes that which pollutes, that which is dangerous and – possibly most important of all – that which can only be organised in units so big and important that they can easily be controlled by small numbers of people . . . "

This movement has then set its face against any further economic growth; against which it enlists ethical concerns and asks us to consider

> "the possibility that some material standard might be regarded (possibly for 'spiritual' reasons) as 'enough' "

but mankind is imperfect and

> "unfortunately experience shows that all countries, however high their levels of material consumption, strive to increase them."

But in the end profligate society will be brought to heel[16]

> "Physical constraints on raw materials as well as energy limit the life of this policy."

The environmentalists advocate a reduction in the use of energy to a level that can be met by the natural alternative energy sources: the physical limits to the world's fossil fuel resources and the probability of an energy gap in oil supplies by the turn of the century will compel the adoption of zero growth, low energy, policies. Nuclear power upsets this plan by offering a new source of energy which can provide society with the means of escape from the stranglehold of limited resources and the environmentalists can no longer rely on their policies being enforced by necessity. Nuclear power must then be opposed at all costs. This is explicitly the message of the Friends of the Earth[17]

> "The nuclear threat is the hardest test the environment movement has yet faced. Nuclear power lies at the heart of a vision of the future committed to an expansion of the present pattern of economic and industrial development. If we wish to argue for alternative patterns of economic and industrial activity, more nationally secure, more personally satisfying and more environmentally sustainable, then we *must* succeed in stopping the development of nuclear power."

This explains the fury directed in particular against the fast reactor. Thermal reactors might even be tolerated since uranium supplies are finite (even though the thorium cycle could extend their life). But the very much larger, almost limitless, energy supply offered by the fast reactor makes this the prime target for attack, hence the rising of concern over plutonium and the problems of proliferation. The US Ford Mitre report is prepared to accept uranium fuelled light water reactors but not the fast breeder.

From this analysis it can be seen that the fundamental issue is a political debate on social values and on what constitutes a desirable future for society: nuclear power is a secondary factor.[18]

> "When environmentalists protest about reprocessing nuclear fuel or pollution they are protesting about a society which values economic and material goals more than quality of life and environmental protection. Nuclear power stations have for them come to have a deep symbolic significance. Their opposition stems from anxieties which go beyond technical questions of risk and safety. Above all they are rooted in growing objections to large, remote, impersonal bureaucracies, increasing dependence on expert elites and reduced participation in the decisions which profoundly affect our lives.
>
> By contrast the supporters of nuclear energy believe in a society dedicated above all to the production of wealth, in which efficiency, cost effectiveness and the needs of industry are the touchstones of policy. If the environment takes a knock or two, or if society takes some calculated risks, then this is the price we pay for the pursuit of the greater good."

This presentation of the "caring" environmentalists in contrast to the materialistic supporter of nuclear power is, of course, a typically disparaging simplification.

The political or even revolutionary aspects of the nuclear opposition should not be underestimated. Many of the leaders of nuclear protest groups have made no attempt to hide their wish to change society. A representative of the Australian anti-nuclear coalition (described as a grass roots group consisting of labour party members, trade unions, church officials and educators) has spoken of being involved in a process for radical social change.[19] The philosophy of the Ecological Movement in France has been well described and analysed by M. Taccoen of Electricité de France.[20] The key idea is that

> "the ideal society should be formed of small self-governing communities living in harmony with nature. This is in opposition to a world of large industrial companies. These are the enemy to overcome. Blindly following their own interests they destroy nature and eliminate mankind. The struggle against nuclear power stations, the hated symbols of the powerful, hierarchical industrial civilisation, is one of the priorities of the Ecological Movement".

The logic of this philosophy requires that the opposition should be extended to the central generation of electricity itself. And this view has been very clearly expressed by Amory B. Lovins, the American who is the British representative of Friends of the Earth Inc.[14]

> "In an electrical world, your lifeline comes not from an understandable technology run by people you know who are at your own social level, but rather from an alien, remote and perhaps humiliatingly uncontrollable technology run by a faraway, bureaucratized, technical elite who have probably never heard of you. Decisions about who shall have how much energy at what price also become centralized – a politically dangerous trend because it divides those who use

energy from those who supply and regulate it."

In view of the New York blackout of 13 July 1977 it is perhaps fair to quote also the succeeding paragraph, if only to show the plausibility of the argument[14]

"The scale and complexity of centralized grids not only make them politically inaccessible to the poor and weak, but also increase the likelihood and size of malfunctions, mistakes and deliberate disruptions. A small fault or a few discontented people become able to turn off a country. Even a single rifleman can probably black out a typical city instantaneously. Societies may therefore be tempted to discourage disruption through stringent controls akin to a garrison state. In times of social stress, when grids become a likely target for dissidents, the sector may be paramilitarized and further isolated from grass-roots politics."

Nuclear power must according to Lovins be rejected because it:[14]

"also has unique sociopolitical side-effects arising from the impact of human fallibility and malice on the persistently toxic and explosive materials in the fuel cycle. For example, discouraging nuclear violence and coercion requires some abrogation of civil liberties; guarding long-lived wastes against geological or social contingencies implies some form of hierarchical social rigidity or homogeneity to insulate the technological priesthood from social turbulence; and making political decisions about nuclear hazards which are compulsory, remote from social experience, disputed, unknown, or unknowable, may tempt governments to bypass democratic decision in favour of elitist technocracy".

The alternative offered is what Lovins calls the "soft path". This treats energy not as an end in itself but only as a means to social ends:

"using our best technologies to wring as much function as possible out of each unit of energy we use".

With this there should be a rapid development of "soft" technologies:

"these are diverse relatively simple technologies that rely on natural energy flows (sun, wind, vegetation) and that are matched in scale and in energy quality to end-use needs".

"These technologies would suffice even in the most industrialised countries, such as those of Europe, Japan, and North America, to meet virtually all energy needs in about 50 years, with no technical breakthroughs whatever, but only using devices that are already demonstrated and already economically attractive. In this sense countries like Japan, Denmark, France and India, though poor in *fuels* are rich in *energy*. A substantial fraction – upwards of a third of all energy needs in the US, for example – could be met with present soft technologies by about the turn of the century. In developing countries simple technologies, oriented toward real human needs, could have a profound impact even more quickly."

The "soft path" is described in lyrical terms as a universal panacea:[21]

> "I do not claim that adopting a soft path will be easy; only easier than not doing it. But properly handled, it can have enormous political appeal. A soft path offers advantages for everyone. It gives us jobs for the unemployed, capital for the business people, opportunities for big business to recycle itself and for small business to innovate, environmental protection for conservationists, better national security for the military, exciting technologies for the secular, a rebirth of spiritual values for the religious, world order and equity for globalists, energy independence for isolationists, racial reforms for the young, traditional virtues of thrift and craftmanship for the old, civil rights for political liberals, and local autonomy for political conservatives. Though it runs against the perceived short-term interests of some powerful institutions, it also runs with many convergent strands of social change at the grassroots.
>
> It runs with, not across, our political grain.
>
> The basic issues of energy strategy, far from being too complex and technical for ordinary people to understand, are on the contrary too simple and political for most experts to understand. It is these simple yet powerful ideas that we want now to carry both to the experts and to people in the village square."

These statements are significant in view of the leading part which the FOE plays in coordinating an international network of protest groups, and in providing them with basic information and arguments on nuclear issues. In France, for example, les Amis de la Terre are considered as "l'expression pure et dure du mouvement écologique" of which they provide the core. It is then worth considering a further quotation from Lovins.

> "For all these reasons, if nuclear power were clean, safe, economic, assured of ample fuel, and socially benign *per se*, it would still be unattractive because of the political implications of the kind of energy economy it would lock us into."

It is important to realise from this that there are no arguments that can convince the opponents: their minds are closed – it is the dialogue of the deaf. Pushed to the extreme, these views now form the basis of a new social revolution. This is spelt out clearly in an article by the French anti-nuclear sociologist Alain Touraine (Ref. 22):

> "A great effort has gone into converting the rejection of the values of industrial society and the fear of a nuclear accident or contamination into a political and social movement. Technocratic power is being fought . . . in the name of a defence of the environment . . . "

Touraine would bring Karl Marx up to date. For the present society the class struggle must take a new form. The enemy is no longer capitalism but technocracy and with the "embourgeoisment" of the workers, they have been replaced by the new social movements: ecologists, regionalists, consumers, feminists etc. "Political parties and trade unions, in particular, still remain attached to older types of political aims and demands." "The new movement is critical of labour ideology, just as the labour movement was critical of bourgeois freedoms. In France and West Germany, certainly, it sees itself both as successor to, and the opponent of, the labour movement." It is "a new revolutionary movement – developing within and desperately trying to pump life into, an earlier revolution that has grown cold and died". "It does not believe that the aim is to help industrial or capitalist society to give birth to a more advanced *socialist* society. It is fighting industrialised society, whether capitalist *or* socialist, in the name of the demand to live differently – now." Touraine shows how the "utopian" wing of the movement whose enemy is Power – "omnipresent, diabolical, a new image of death" – can make use of local campaigns in which action outweighs utopianism as with a local community defending itself against a nuclear power plant. This merges into a populist form of action where the anti-nuclear campaign is often linked with regional movements. This can be clearly seen in Spain where the Basque separatist movement has launched a number of bomb attacks against the Lemonitz power station. In the United Kingdom, a country where nuclear opposition is relatively weak, the main protest movement is associated with the Scottish Nationalists.

Nuclear power then becomes the lowest common denominator of all protest movements: environmentalists concerned about the protection of nature and the countryside; pacifists against war and the spread of nuclear weapons; separatists and home rule groups against the central authority of the state; advocates of zero growth denouncing the waste of energy and natural resources. It unites all who are pessimistic of the future and whose bitterness against the modern world is expressed as a nostalgia for an idealised past or a utopian dream of a more "natural", "alternative" way of life. But to quote Touraine once more:

> "Energy policy is only one of the principal areas of technocratic domination. Changed circumstances may make some other areas of social life, or a different type of protest, come to prominence. Ecologists are perhaps already looking for other battlefields, even though the battles over nuclear sites are not over."

It is then tempting, against this background, to try to locate the anti-nuclear movement within a recent analysis of the style of political action.[23] This study identified five types of political involvement: inactives, conformists, reformists, activists and protesters. Of these the reformists and activists while differing in the extent to which they would carry their levels of protest share an involvement in the conventional political system. The activists tend to be young, with political leanings towards the "left" and hold post-materialist values in favouring "a society where ideas are more important than money". They are highly competent, efficacious, politically conscious – and disaffected. The protesters on the other hand tend to remain outside politics and have a low level of political interest; they are mainly young, but surprisingly were found to have a dominant proportion of women. "The protester category can be partially described as a curious alliance between poorly educated young men and some well educated young women." While the protesters have a commitment to action and would take part in demonstrations, some to the extent of illegal acts of protest, they exhibit no particular political involvement. Their commitment to action is associated with a lack of commitment to anything else.

> "So what are they likely to protest about? The answer is that it will depend on who gets to them first . . . The main source of unease is the knowledge that there have been times in recent history when the protesters have been easily mobilised by groups of determined 'activists' whose intentions have been anything but democratic."

It may then be significant that, according to Dr Marsh, the percentage of protesters ranges from about 20% in the USA and UK, up to 30% in the Netherlands, and if the activists are also included, the two groups account for between 32.6% (UK) and 51.2% (Netherlands) of the population – figures which are close to those shown in many public opinion polls to be opposed to nuclear power. However, where the environmentalists have come forward as political groups as the "Green" or "Ecological" parties, and have submitted their views to the electorate, they have only received a few per cent of the total votes. The difference between this small percentage of committed support for environmentalist policies and the 30–40% of more general opposition must be an indication of the extent to which the underlying fears have been utilised to take a hold on the public mind.

REFERENCES

1. BEIR Committee, 1972.
2. P. Taylor, "Proof of Evidence on Behalf of the Political Economy Research Group", Windscale Inquiry.
3. Joint Committee on Atomic Energy, US Congress, January 1974.

4. W. Patterson, "Transcript of Proceedings: Day 53", Windscale Inquiry.
5. *WHO Technical Report Series*, No. 151 (1958).
6. J. Francis, *A History of the British Railways – Its Social Relations and Revelations*, 1851.
7. G. Eliott, *Middlemarch*, 1872, Chapter 56.
8. H. Marcuse, *One Dimensional Man*, Routledge and Kegan Paul.
9. *Futures*, 8 (No. 1), (February 1976).
10. M. Tubiana, Colloquium on the Psycho-Sociological Implications of the Nuclear Industry, Paris, January 1977.
11. M. Tubiana, *Revue Generale Nucleaire*, No. 5 (1979).
12. Hoyle, *Energy or Extinction*, Heinemann Educational Books, London.
13. Vassily Yemelianov, USSR Academy of Sciences, quoted in *Eastern Business Magazine*, 1977, p. 241.
14. A.B. Lovins, "Energy Strategy. The Road Not Taken", *Foreign Affairs*, (October 1976).
15. Lord Beaumont of Whitley, "The Green Alliance", letter in *Financial Times*, 4 October 1978.
16. J. Davoll, letter to *Atom*, No. 264 (October 1978).
17. T. Burke, letter to *Vole*, October 1978.
18. S. Cotgrove, letter in *The Times*, 27 November 1978.
19. Report on Conference for a Non-nuclear Future, Salzburg, April 1977.
20. L.B.C. Taccoen, "Movements Anti-nucleaires, Opinion Publique et Politique en France", IAEA-CN-36/FR 255.
21. A.B. Lovins, Conference for a Non-nuclear Future, Salzburg, April 1977.
22. A. Touraine, *New Society*, 8 November 1979.
23. A. Marsh, "The New Matrix of Political Action", *Futures*, April 1979.

18
Conclusion

The argument of this book is that the inevitable growth of world population, together with the desire of the major part of that population which lives in developing countries to have a higher standard of living, approaching more closely that of developed countries, will push world energy demand to between two to three times the present level by the year 2000 – a short twenty years away. To meet this demand by that time the only substantial energy sources are oil, coal and nuclear power. Oil and gas output is unlikely to rise much, if at all, above present levels and could even fall. There must then be a greater reliance on coal. Coal production could be expanded, but it will take time to establish new mines and in particular the new international coal trade that will be required for coal to make a much larger contribution to world energy needs. Coal will be increasingly used as a source of hydrocarbon fuel to substitute for oil leaving uranium to take over as the main primary fuel for electricity generation, and in the longer term for the production of hydrogen. Nuclear power will also be used as a source of heat. Those countries in a position to utilise nuclear energy should then increase their programmes to the greatest extent possible to reduce the present dependence on oil and relieve the pressures on oil supply. And experience shows that nuclear power is cheaper, safer, and less environmentally polluting than most other sources of energy.

These arguments are, however, far from being universally accepted (otherwise this book would not have been written) and where there are differences of view on future needs it must be recognised that no one can claim to be infallible. With all the uncertainties in attempting to foresee the course of events the assumptions and predictions made here may prove to be erroneous. Perhaps world population growth can be halted, quickly; perhaps extreme conservation measures will bring energy demand down to lower levels consistent with a stationary energy supply; perhaps the amount of energy that might come from wind, waves, the sun etc. has been grossly underestimated; perhaps there will be a sudden breakthrough in the technology of these 'renewable' energies to enable them to supply substantial amounts of energy at an economic cost; perhaps society, and particularly the developing countries, will turn away from the path of industrial development and seek an alternative low energy life-style. Rather than continue to argue over the probability that one or the other view of these events may be more realistic it is more useful to consider the consequences of being wrong.

Let us suppose the recommendations of this book are followed and over the next twenty years a substantial number of new nuclear power stations are ordered and will be in operation or under construction. Most of this new nuclear plant will necessarily be built in those countries which have already started on the path of nuclear development. But if by the turn of the century it becomes clear that nuclear power is not needed the only consequence will be that the number of operating stations will have been increased by perhaps a factor of between 2–5, with a larger number of stations partly under construction or being planned. This will not greatly increase the world stock of radioactive materials or create significantly greater problems should it be decided to close down and dismantle the operating nuclear stations, nor will it alter the potential of any of the countries to develop a nuclear weapon capability that they do not already have.* But the availability of a new energy source means that there need be no major or dramatic changes from the present way of life.

Suppose on the other hand that the arguments put forward by those who oppose nuclear power are adopted, then either a new and, as yet unknown or unproven, energy source must be found and introduced on a large scale over the next twenty years or the increase in the world's population must be halted. Unless this can be done the world per capita consumption of energy will be sharply reduced or else the gap between the high energy consumption of the industrial countries and the low energy consumption of the developing countries will widen.† The consequences of either course are likely to impose severe strains upon society. The enforced acceptance of zero or negative economic growth which would follow a reduction in per capita energy consumption is likely to lead to economic recession and increasing unemployment or to a labour intensive society or even both at the same time.

* This last point is underlined by the reports that China, now facing serious energy shortages, may at last take up the development of nuclear power – some sixteen years after China became the fourth country in the world to have exploded an atom bomb.

† "The present inequality in the consumption of commercial energy is strikingly presented in the report of the 'Brandt Commission' which points out that one American uses as much as 3 Swiss, 9 Mexicans, 53 Indians and 1072 Nepalese." (North-South: A programme for survival)

The Brandt Commission in their 1980 report "North-South: A programme for survival" warns of the major tensions that will arise between countries competing for energy and underlines "how essential it is that . . . the world's energy problems be solved by peaceful means". And indeed the danger that competition over scarce energy reserves may lead to war is far from negligible. Mr James Callaghan, the former UK Prime Minister, has warned . . . "on oil, the struggle is going to be so fierce and intense because this source of energy is going to run out, that in twenty years time unless we have found alternative sources of energy, the world could be at war over the Middle East".*

Time will in the end show what fundamental changes in the pattern of world energy production and consumption will occur. The acceptance of nuclear power would ensure that any transition can be softened by the assurance that a new energy source is available. The consequence, should it not be needed, will be minimal. But the consequence of a world plunged into severe and widespread energy shortages would be catastrophic. This is a risk the world cannot afford to take. Nuclear power is no longer an option, it is a necessity that the world cannot forego.

* BBC-1 Television programme, 1 November 1979.

Glossary

Actinides: Heavy elements with an atomic number greater than 89. The series includes, actinium, thorium, protactinium, uranium, neptunium, plutonium, americium, curium, berkelium and californium, all of which are chemically very similar. Those of interest are the long-half life alpha-emitters.

Alpha particle: A heavy particle produced by a radioactive decay process and in various nuclear reactions.

Back-end of the fuel cycle: Those fuel cycle processes concerned with the treatment of spent fuel discharged from reactors.

Background radiation: The natural ionising radiation of man's environment including cosmic rays from outer space, naturally radioactive elements in the ground, and naturally radioactive elements in a person's body. The average radiation dose from natural radiation is of the order of 100 millirem/year.

Boiling Water Reactor (BWR): A nuclear power reactor cooled and moderated by light water. The water is allowed to boil in the core to generate steam which passes directly to the turbine.

Breeder reactor: A type of reactor in which the production of fissile nuclei (plutonium-239) in a blanket of fertile uranium-238 surrounding the core exceeds the consumption of fissile nuclei (plutonium-239) and uranium-238) in the core.

Burn-up: A measure of the quantity of energy that has been obtained from a sample of nuclear fuel in a reactor core. It is usually measured in units of megawatt-days per tonne of fuel.

Centrally planned economies: A grouping of countries used by the United Nations. Countries included are the USSR, eastern Europe, China, Mongolia, North Korea and Vietnam.

CMEA: Council for Mutual Economic Assistance. USSR, German Democratic Republic, Bulgaria, Hungary, Czechoslovakia, Roumania, Yugoslavia, Poland, and Cuba.

Coal and oil equivalents: 1 tonne coal = 0.6 tonne of oil; 1 tonne oil = 1.67 tonnes of coal.

Containment: The structures, within and including the reactor building, designed to prevent any material that may escape from the reactor itself from reaching the outside environment. The reactor containment usually consists of steel and thick concrete.

Contamination: Radioactivity where it should not be.

Control rod: Rod of neutron-absorbing material inserted into reactor core to soak up neutrons and shut off or reduce rate of fission reaction.

Cooling ponds: The water-filled tanks at a power station or reprocessing plant in which spent fuel rods are placed after removal from the reactor.

Coolant: A liquid or gas circulated through the core of a reactor to extract the heat of a fission process, to provide shielding and to remove the decay heat.

Core: The central part of a nuclear reactor containing the fuel rods, moderator and control rods. Nuclear fission reactions take place and heat is generated within the core.

Cosmic radiation: Electrons and the nuclei of atoms, largely hydrogen, that impinge upon the earth from all directions in space with nearly the speed of light.

Critical: Refers to a chain reaction in which the total number of neutrons in one 'generation' of a chain reaction is the same as the total number of neutrons in the next 'generation' of the chain; that is, a system in which the neutron density is neither increasing nor decreasing.

Critical mass: The minimum amount of fissile material needed to sustain a chain reaction.

Curie (Ci): The unit of radioactivity, defined as 37,000 million (3.7×10^{10}) disintegrations per second. This is almost equal to the radioactivity of 1 g of radium-226. Quantities of radioactive material are commonly measured in curies; one curie of material is the quantity having an activity of one curie. 1 curie = 3.7×10^{10} Bequerel (Bq).

Daughter/Daughter product: The substance into which a radioactive nucleus transforms itself by radioactive decay.

Decay: The decrease in activity of a radioactive material as it spontaneously transforms from one nuclide to another or into a different energy state of the same nuclide.

Decay heat: Heat generated by radioactive disintegrations which occur in irradiated fuel. This is additional to heat from chain reaction; it cannot be shut off but decays slowly with time.

Decontamination: Transfer of unwanted radioactivity to a less undesirable location.

Depleted uranium: Uranium in which there is less fissile isotope uranium-235 than the 0.71% normally found in natural uranium.

Deuterium: An isotope of hydrogen containing one proton and one neutron in the nucleus. Chemically similar to hydrogen, but of twice the mass and with different physical and nuclear properties. Its natural abundance is about one part in 7,000 of hydrogen.

Developed countries: A grouping of countries used by the United Nations. They include the USA, Canada, western Europe including Yugoslavia and Greece, Israel, South Africa, Japan, Australia and New Zealand.

Developing countries: All the countries of the world other than he developed countries and countries with centrally planned economies. This group includes the countries of Central and South America, and most of the countries of Africa and Asia.

Dose: Amount of energy delivered to a unit mass of a material by radiation travelling through it.

Dose-rate: Time rate at which radiation delivers energy to unit mass of a material.

Doubling time: The time required to produce as much additional plutonium as was required for the total (in-reactor and out-of-reactor) inventory of the initial plant.

Enriched uranium: Uranium in which the content of the fissile isotope, uranium-235, is higher than the 0.71% normally found in nature. Low enriched uranium, containing 2–4% of uranium-235, is used as fuel in many types of reactor. High enriched uranium, which may contain more than 90% uranium-235, is used as fuel in some types of reactor and also to make nuclear weapons.

Fast reactor: A type of nuclear reactor in which the concentration of fissile nuclei in the fuel is high enough for the nuclear reaction to be sustained by fast (i.e. unmoderated) neutrons.

Fertile: As applied to a nuclide, it means capable of absorbing (capturing) a neutron to be transformed into a fissile nuclide. Uranium-238 and thorium-232 are fertile isotopes, giving rise to plutonium-239 and uranium-233 respectively.

Fissile: Capable of undergoing fission.

Fission: The process by which a nucleus splits into two approximately equal fragments plus several free neutrons, giving off large amounts of energy which appears as gamma radiation and heat. Fission occurs spontaneously in certain heavy elements, but in a chain reaction it occurs when a fissile nucleus absorbs a neutron.

Fission products: The mix of nuclides resulting from fission. Fission products are often unstable and undergo radioactive decay, emitting radiation of different types and energies.

Fossil fuel: Coal, oil, natural gas or other carbon-containing fuels derived from the fossilisation of living matter (mostly plants) that flourished hundreds of millions of years ago.

Front-end of the fuel cycle: All those fuel cycle processes from the prospecting for uranium to the preparation of uranium fuel for insertion in reactors.

Fuel bundle: An assembly of metal tubes containing nuclear fuel pellets ready for insertion in a reactor.

Fuel pellets: Uranium dioxide, or other nuclear fuel in a powdered form, which has been pressed, sintered and ground to a cylindrical shape for insertion into the sheathing tubes of the fuel bundle.

Fusion: Short for thermonuclear fusion, a type of nuclear reaction in which two light nuclei, such as deuterium, fuse together to form one heavier nucleus, in the process releasing a large amount of energy. Fusion reactions only take place at exceedingly high temperatures. They are the source of the energy given off by the sun and also of most of the energy released when a hydrogen bomb explodes.

Gamma ray: A form of electromagnetic radiation, similar to light or X-rays, distinguished by its high energy, high penetrating power and short wavelength. Gamma radiation is emitted from many nuclei when they are undergoing radioactive decay and in many other nuclear reactions.

Gas-cooled reactor: A nuclear reactor in which a gas, such as carbon dioxide or helium is used as the coolant.

Half-life: The time taken for half the atoms of a radioactive substance to disintegrate; hence the time to lose half its radioactive strength. Each radionuclide has a unique half life ranging from a millionth of a second to billions of years. The longer the half-life the lower the rate of radioactive decay; atoms with very long half-lives approximate to stable elements for which the half-life is infinity.

Heavy water: Water in which the hydrogen is replaced by its isotope deuterium. It is sometimes called deuterium oxide and occurs in natural water to the extent of about 1 part in 7000. It is the most effective neutron moderator available for reactors.

High level waste: The most highly radioactive waste from fuel reprocessing containing most of the fission products from spent fuel and typically containing millions of curies per cubic metre when first separated. It also contains small amounts of unseparated uranium and plutonium plus the greater proportion of the other actinides produced in the reactor.

HTGR: High temperature gas cooled reactor.

Intermediate level waste: A somewhat arbitrary classification of part of the waste from fuel reprocessing and the nuclear cycle typically containing thousands of curies per cubic metre.

Irradiated fuel: Nuclear fuel that has been used in a nuclear reactor.

Isotope: Atoms of an element having the same number of protons in their nuclei but different numbers of neutrons. All have the same chemical properties as the element and thus cannot be separated by chemical means. However, they can be separated by using certain physical processes, such as gaseous diffusion.

Joule: The unit of energy or work in the metre-kilogram-second system of units. It is the work done by a force of one newton when the point at which the force is applied is displaced one metre in the direction of the force.

Light water: Ordinary water with normal hydrogen atoms, as distinct from heavy water.

Low level waste: Part of the waste from various stages of the nuclear fuel cycle typically containing a few curies per cubic metre.

LWR: Light water reactor.

Magnox: A type of magnesium alloy used as cladding for the metallic fuel in early British and French gas-cooled reactors. In Britain, the word has been adopted as the name of the reactor type.

Metrix multipliers:

k (kilo) = 10^3	m (milli) = 10^{-3}
M (mega) = 10^6	μ (micro) = 10^{-6}
G (giga) = 10^9	n (nano) = 10^{-9}
T (tera) = 10^{12}	p (pico) = 10^{-12}
E (exa) = 10^{18}	

Mixed oxide: Reactor fuel in which the fissile nuclei are plutonium-239, mixed with natural or depleted uranium. These elements are used in oxide form.

Moderator: A material used in a reactor core to slow down fast neutrons, without unduly absorbing them, so as to increase the probability of the neutrons causing fission in a uranium-235 or plutonium-239 nucleus.

MTCE: Million tonnes coal equivalent.

MTOE: Million tonnes oil equivalent.

MW(e), MW(th): One megawatt (MW) is a unit of power equal to one thousand kilowatts. MW(th) denotes the thermal power of a power station, that is the rate at which heat is produced. MW(e) denotes the electrical power output of the station and is a fraction of the thermal power – typically about 33% for a light water reactor and up to 40% for a modern fossil fuel-fired power station, MTGR or fast reactor. This ratio is called the thermal efficiency of the power station.

Natural uranium: Uranium whose isotopic composition as it occurs in nature has not been altered (0.71% by weight of U-235).

Neutron: An uncharged particle which is a constituent of the nucleus of all nuclides except hydrogen; neutrons are ejected from the nucleus in nuclear reactions, such as fission.

NPT (Non-proliferation treaty): Intended to control the spread of nuclear weapons and their technology.

Nuclear energy: The energy liberated by a nuclear reaction such as fission.

Nucleus: The positively charged core of an atom which has almost the whole mass of the atom but only a minute part of its volume. All nuclei are made up of protons and neutrons, except for ordinary hydrogen (H) which contains only one proton.

Nuclide: A nuclear species, all the atoms of which contain similar nuclei.

OECD: The Organisation for Economic Co-operation and Development with headquarters in Paris. Its member states are: Australia, Austria, Belgium, Canada, Denmark, Federal Republic of Germany, Finland, France, Greece, Iceland, Ireland, Italy, Japan, Luxembourg, Netherlands, New Zealand, Norway, Portugal, Spain, Sweden, Switzerland, Turkey, U.K. and U.S.A.

Plutonium: Heavy artificial metal, made by neutron bombardment of uranium; fissile, reactive chemically, toxic, alpha-emitter.

Pressurized Water Reactor (PWR): A power reactor cooled and moderated by light water in a pressure vessel surrounding the core. The water is pressurized to prevent boiling in a closed primary loop and is circulated through a heat exchanger which generates steam in a secondary loop connected to the turbine.

Pressure vessel: In most types of reactor, including LWR's and gas cooled reactors, the pressurised coolant is confined in a single large vessel made of welded steel or concrete. This is called a pressure vessel, and the whole of the reactor core is contained within it.

Proton: A positively charged particle which is a constituent of the nucleus of all nuclides. The number of protons in the nucleus determines the chemical properties of an element and hence is characteristic of each of the chemical elements.

Rad: The unit dose of ionizing radiation. One rad is absorbed when 1000 ergs of energy is imparted to each gram of matter by ionizing radiation (see Rem). 1 rad = 1/100 of a Gray (Gy).

Radioactivity: The spontaneous decay of an unstable atomic nucleus into one or more different elements or isotopes. It involves the emission of particles by spontaneous fission until a stable state is reached.

Reactivity: A measure of the departure of a reactor from criticality. A positive value means that the release of neutrons is increasing and the power will rise.

Recycling: The re-use as a reactor fuel of uranium and plutonium extracted from spent fuel elements.

Rem: Roentgen equivalent man, the unit of dose equivalent of ionizing radiation in biological matter. It is the absorbed dose in Rads multiplied by a factor which takes into account the biological effect of the radiation. 1 rem = 1/100 of a Sievert.

Reprocessing: The stage of the nuclear fuel cycle at which plutonium and uranium in spent fuel are recovered from the other actinides and the fission products, which constitute waste.

Safeguards: A series of international arrangements designed to detect and hence deter the use of nuclear facilities or materials for prohibited purposes, such as the production of nuclear weapons.

Separative Work: A measure of the work done in enriching uranium from the initial value to the desired final enrichment. SWU, separative work units in kilograms of separative work.

Shielding: A mass of material that reduces radiation intensity to protect personnel, equipment or nuclear experiments from radiation injury, damage or interference.

Spent fuel: Nuclear fuel that has been irradiated in a reactor to the extent that it can no longer effectively sustain a chain reactor; the fissionable isotopes have been consumed and fission-product poisons have accumulated.

Tailings: The finely ground waste material from an uranium mill after the uranium has been extracted from the ore.

Tails: The depleted uranium produced at an enrichment plant.

Tails assay: The proportion of uranium-235 remaining in the tails after enrichment. It is typically between 0.2 and 0.3%. The tails assay is an important characteristic of the operation of an enrichment plant; the higher the tails assay, the less will be the amount of enrichment energy to produce a given quantity of enriched uranium, but the greater will be the amount of natural uranium feed required.

Thermal: Of neutrons, that are travelling at a relatively slow speed, comparable with that of gas molecules at ordinary temperatures.

Thermal reactor: A type of nuclear reactor in which most of the nuclear fissions are caused by thermal neutrons that have been slowed down by a moderator. Most power reactors at present are thermal reactors. They consume more fissile material in total, in the form of uranium-235 and plutonium-239 bred in the reactor, than they produce in the form of plutonium-239.

Thorium: A slightly radioactive metallic element with an atomic number of 90, whose naturally occurring isotope Th-232 is fertile and the source, when irradiated in a reactor, of U-233.

Ton: 1 ton = 2240 pounds.

Tonne: Metric ton; 1 metric ton = 2204.6 pounds.

Uranium: Heaviest natural element, dark grey metal; isotopes 233 and 235 are fissile, 238 fertile; alpha-emitter.

Uranium enrichment: The process in which the ratio of the concentration of the isotope uranium-235 to that of the isotope uranium-238 is increased above that found in natural uranium.

Uranium milling: The process in which uranium is recovered from ore.

U_3O_8: The formula for uranium oxide in the form in which it is contained in yellowcake.

Working level: The quantity of radon-222 decay products in one litre of air which will result in the ultimate emission by them of 1.3×10^5 million electron volts of alpha particle energy. If the short-lived decay products are in equilibrium with the radon in air, then 100×10^{-12} curies per litre of radon is equivalent to one Working Level.

Working Level Month (WLM): Exposure to radon decay products at a concentration of 1 Working Level throughout the working period of a month (defined as 170 hours).

Yellowcake: The mixture of uranium oxides and impurities (typically about 95 per cent U_3O_8) produced at a uranium mill.

Zircaloy: Alloy of zirconium used as fuel cladding; has low cross-section for absorption of neutrons.